AF883613

iPad® Picture Book for the Absolute Beginner

Just the Basics for Seniors

Wendy S. Weston

iPad® Picture Book for the Absolute Beginner
Just the Basics for Seniors

Copyright © 2018 by Wendy S. Weston.
All rights reserved. Except as permitted under the Copyright Act of 1976, no part of this publication may be reproduced or distributed in any form or by any means, or stored in a database or retrieval system, without the prior written permission of the publisher. No patent liability is assumed with respect to the use of the information contained herein.

Trademarks
iPad® is a registered trademark of Apple Computer, Inc. All other trademarks are acknowledged as belonging to their respective companies, and any references to such trademarks or logos are for demonstrative or descriptive purposes only.

Limit of Liability/Disclaimer of Warranty:
Every effort has been made to make this book as accurate as possible. This book does not cover every single aspect, nor was it intended to do so. Instead, this book is meant to cover some absolute basics only. No warranty or fitness is implied. The advice and strategies contained herein may not be suitable for your situation. You should consult with a professional where appropriate. The author/ publisher shall have neither liability nor responsibility to any person or entity with respect to any loss or damages arising from the information contained in this book. The fact that an organization or website is referred to in this work as a citation and/or a potential source of further information does not mean that the publisher and author endorse the information the organization or website may provide or recommendations it may make.

Names and personal information have been changed throughout this book, in order to protect the privacy of individuals.

iPad® Picture Book for the Absolute Beginner - Just the Basics for Seniors
Publisher: Wendy S. Weston
Author: Wendy S. Weston
ISBN: 978-1-7338279-0-4
Printed in the U.S.A.

Foreward
by Aunt Stella (97 years young)

My husband would tell the story about Emperor Franz Joseph, who reigned over the vast Habsburg empire until the First World War. Old Franz was not known as a great brain, but he was quite popular, so my grandparents told me. Anyway, when he was shown the first telephone, he said it was a nice toy. If he could be so shortsighted, we (my husband and I) were too.

For years my nieces and nephews wanted me to get a computer. I said, “what for?” I was told I could look up whatever I wanted to know. But I had an Encyclopedia and a phone book and a dictionary. What else did I need to know?

I also thought I would never understand how to use a computer. I tried a few lessons and gave up more than a few times. I tried to understand the computer for dummies books, but they did not work for me, and I told my family to give up. At my age I did not need to be computer literate.

But my niece, Wendy, would not give up. She had an idea. She bought an iPad for me, and she used photographs of the screen, and showed me step by step what to do. I played around with it and, before I knew it, I got results! First came email, and then Google. And then I did something wrong and “killed” my iPad. Help! I could hardly wait for Wendy to call me back and tell me how to revive my iPad. How could I go just one day without it? Isn’t it funny how much you can miss something you went without for the better part of a century? (I forgot to mention that Wendy lives in another State, and all of my lessons were by her book and some phone calls with her.)

Whenever I look something up on the internet, it leads to something else I want to know. It’s like a toy for oldies! I wake up and look at my email. I read the news on my iPad. And now I read books on it, in large print. I borrow e-books from the library, it’s all free, and I don’t have to physically go there at all.

Understanding the iPad has been such a gift for me. I have been able to reconnect with many long distance friends and family members. I truly enjoy emailing and messaging with them and knowing what’s happening in their lives. Now I also own an iPhone and thanks to my knowledge of the iPad, learning how to use the iPhone was easy!

Someone recently called me the computer guru. Haha! I wish. But I keep learning, and I can’t live without it. I believe you will also feel this way, when you try it. In this day and age we need to become “literate” all over again. I wish I had started 20 years ago, because I missed so much. Instead, I started at the young age of 95. It just goes to show, you’re never too old to learn something new!

From her many years of teaching experience, Wendy has a greater understanding of how to effectively teach beginners. I will be forever grateful to her for bringing me into, and helping me understand more of, the modern world.

Aunt Stella
iPad Student Extraordinaire!

Preface

A few years ago my Aunt Stella, who was 95 years old at the time and who had tried a few times and given up, was ready, at my insistence, to try one last time to learn how to use an iPad. I had finally realized that what she needed, in order to experience success, was an iPad guide which took her step-by-step through the process.

My thirty-seven years of experience in teaching music to very young children (absolute beginners) sparked my enthusiasm for this new challenge. I gave a lot of thought to the best approach for this book:

How did I teach my students?
One step at a time!

How did they learn most effectively?
By demonstration! (Not by reading wordy explanations.)

So I got to work and made a picture book with simple step-by-step instructions and somewhat primitive hand-drawn arrows. (That was before I learned how to create nice clean arrows on my computer!)

I mailed my very rough version of the beginner's book to my aunt. Lo and behold, it worked! Aunt Stella finally understood the basics of using an iPad and she was having fun with it! She urged me to put the book out so that other seniors would benefit as well. And that is how this new book came to be.

I do not consider myself to be a computer expert, far from it! I just understand how most absolute beginners learn.

Acknowledgements

This book would never have been conceived if not for my Aunt Stella who, at the age of 95, provided me with the opportunity to prove myself worthy as her teacher. Thank you, Aunt Stella, for your willingness to give my approach a try and for your encouragement to create this book for all of the other seniors out there who may desire a simpler method for learning about their newfangled devices in this modern world.

I am grateful to my family who, perhaps unknowingly, encouraged me with their praise of my very rough first draft, which was made specifically for Aunt Stella. To all of my family members, you may never know how much your words of praise inspired me to go forward with this book. Thank you.

Special thanks to my husband, Rob, for his love, support and patience during this process. Rob, you were always a willing and enthusiastic participant every time I needed advice and edits. You're my rock and you rock my world!

About This Book

When using this book, be sure to learn the various iPad terms on page one: **Glossary of Terms**, and refer back to it anytime you need a reminder of any definitions.

This book is intended to guide the Absolute Beginner and covers only the basics for using an iPad. Once the basics have been mastered you may want to learn more. If that's the case, there are many books for senior beginners out there which go into much more detail and explore more iPad features than I do here.

Although written with an iPad in mind that does not have a cellular plan, much of the content of this book is still relevant to iPads with cellular plans, except where internet connection is stated as "required." iPads which have cellular plans do not require internet connections in order to perform certain functions.

All iPads are powered by an operating system (iOS). The operating system on the iPad used for demonstrations in this book is iOS 11. All new iPads will come with iOS 11, or higher, already preinstalled. If your iPad is an older model, you can easily upgrade to a more current iOS by following these steps: tap **Settings** on your iPad, then tap **General**, then tap **Software Update**, then scroll down and tap **Install Now**. Your iPad will do the rest for you. The iOS update will take a while to install.

There are many options for receiving various notifications on your iPad. I feel that, for the absolute beginner, such notifications can be quite distracting and confusing. Therefore, you'll notice that in several places in this book, I recommend turning them off for the beginning stages of learning just the basics of your iPad. Notifications can always be turned back on at a later date, once the basics have been mastered. I believe the most effective learning is accomplished one simple step at a time, with as few distractions as possible.

I strongly recommend that you purchase a protective case for your iPad. If accidentally dropped, an iPad's glass screen can crack or other functions can be affected, rendering your iPad unusable. Screen protectors are also available, if you choose to use one. Both of these items can be purchased at your local Apple Store, electronics store, or on the internet.

Table of Contents

Glossary Of Terms

App: Short for "application." The programs you can download and use on your iPad.

Apple ID: An email address registered with Apple.

Cursor: A small, blue vertical line that blinks to show where your text will be entered.

Dock: A row of icons along the bottom of your iPad screen that don't change when you swipe between home screens.

Emoji: A small digital image used to express an emotion or idea.

Facebook: A social network on the internet, built for friend-to-friend connections.

FaceTime: A video and audio calling app for calling people who also have Apple devices.

Home Button: The round button on the bottom of the front of the iPad, beneath the screen.

Home Screen: The page that shows all of the icons for the apps on your iPad.

iMessage: The application for sending and receiving text messages.

Icons: Small pictures which represent applications on your iPad. By touching an icon you will open that particular application.

iOS: An operating system which powers devices manufactured by Apple Inc.

Jiggly Mode: All icons will begin to "jiggle" if an icon has been touched and held for 3 seconds or more. This allows icons to be moved into different positions on the home page. To stop the jiggling, press the Home Button.

Safari: An application for using the internet.

Scroll: Hold finger on the screen and slide it up or down to scroll the current page up or down. (Think of it as moving the page up or down.)

Status Bar: A white strip across the top of the screen which shows information about the iPad and the app that is being used

Swipe: The most basic iPad gesture is swiping your finger to scroll through pages or lists. To scroll down on a page, place the tip of your finger at the bottom of the screen and slide it towards the top of the screen (this is swiping up). At first, it may seem counterintuitive to scroll down by swiping up, but if you think of it as your finger moving the image on the screen, it makes sense. To scroll up on a page, place your finger at the top of the screen and slide it toward the bottom of the screen (this is swiping down).

Thumbnails: Reduced-size versions of photos.

WiFi: Wireless internet connection.

What's In The Box

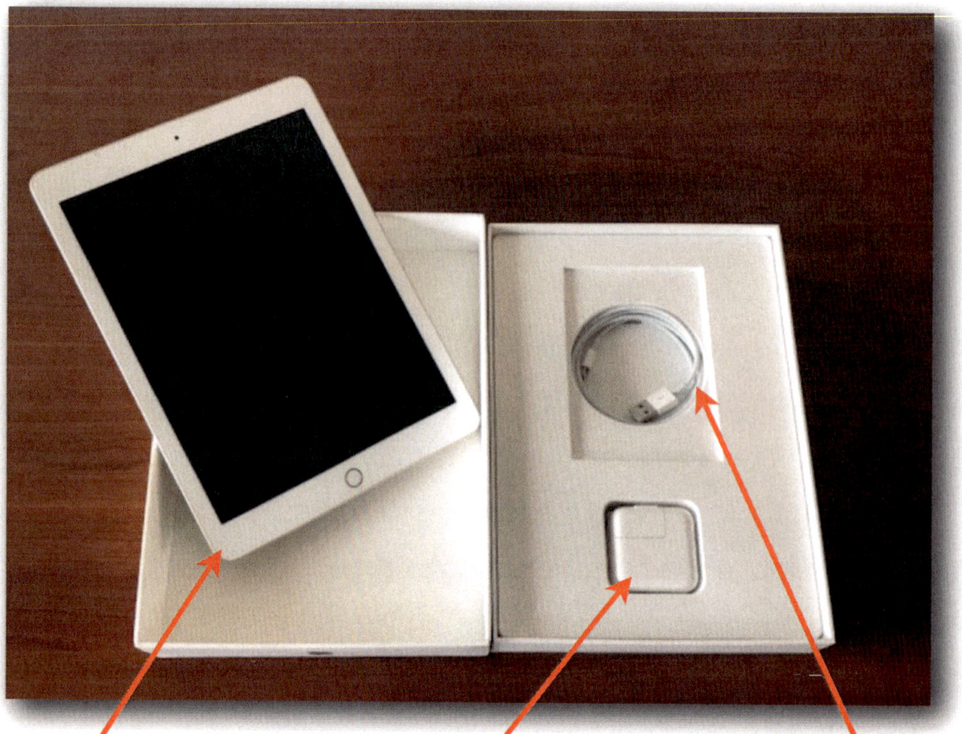

iPad

USB Power Adapter

Lightning to
USB cable
(charging cable)
connects the
iPad to the
power adapter

iPad Diagrams

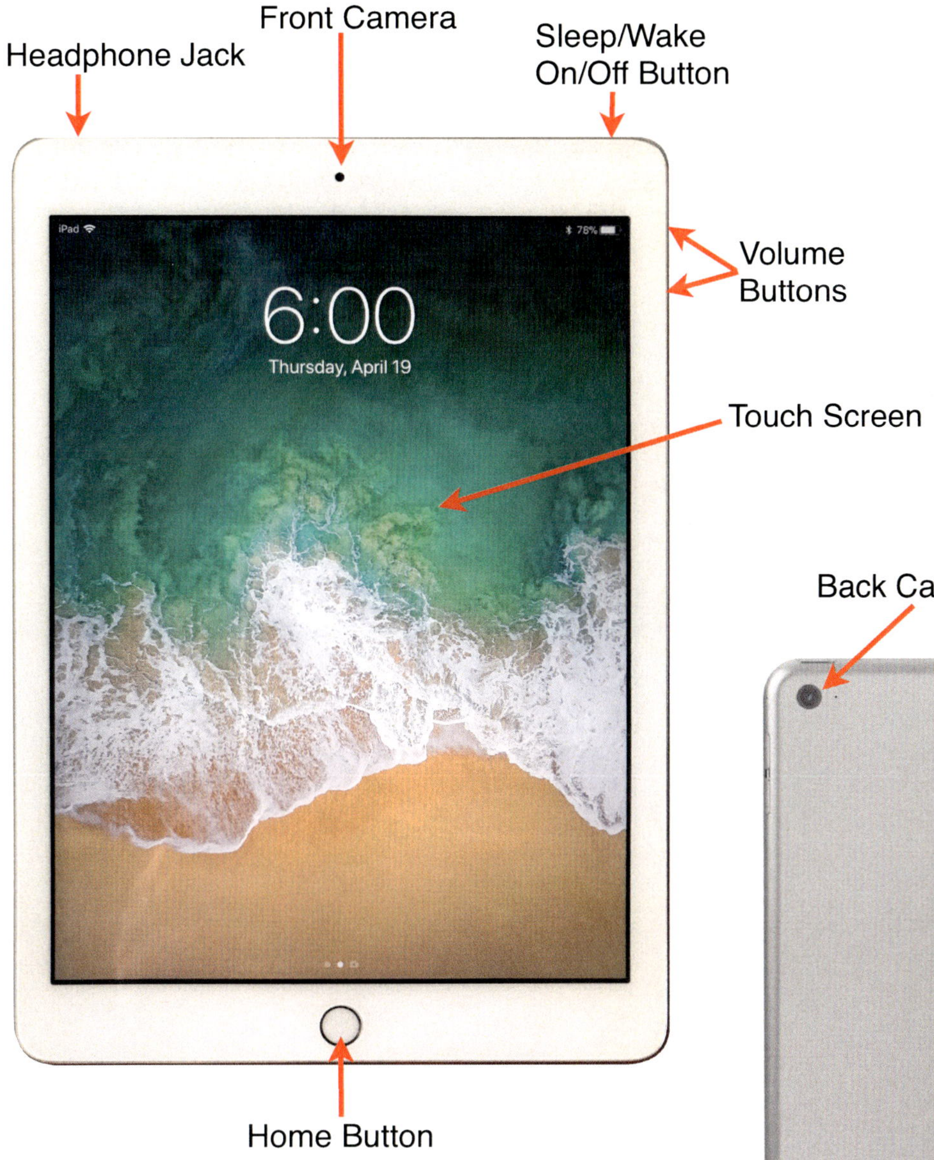

Charging The Battery

When your battery charge is running low your iPad will let you know. At that time you will need to charge your iPad. Some choose to charge their iPad every night, and some, like myself, choose to charge it only when the charge is running low. Either way is fine. The choice is yours.

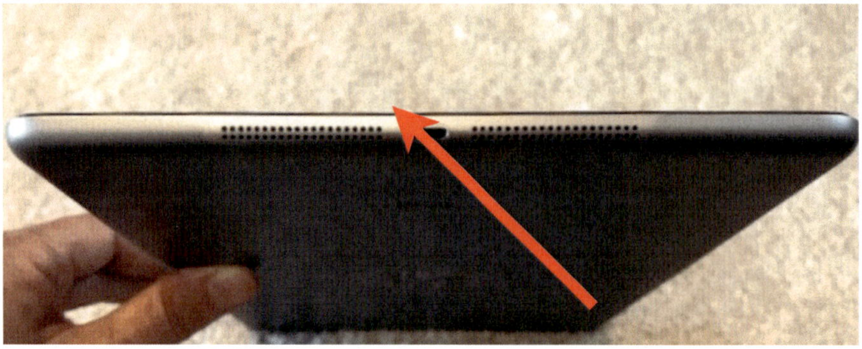

This is the Charging Cable Port

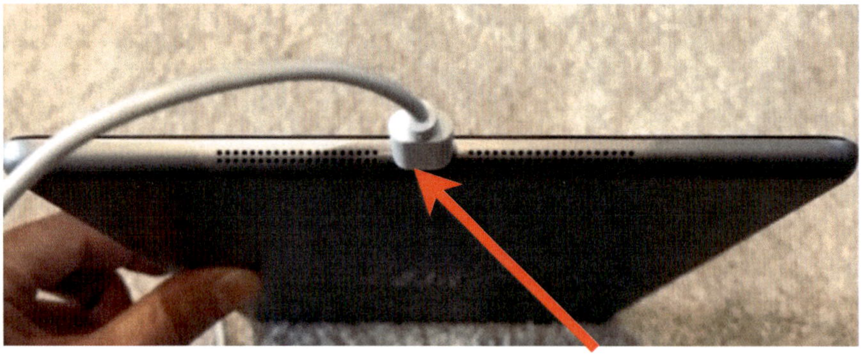

Plug in your Charging Cable here

Battery status is displayed here.

Turning Your iPad On Or Off And Using Sleep Mode

Turning Your iPad On:

Press and hold the Sleep/Wake Button until the Apple logo appears on the screen. Then release the button.
It may take a few minutes for your iPad to turn on completely.

Turning Your iPad Off:

Press and hold the Sleep/Wake Button until you see this on the screen:

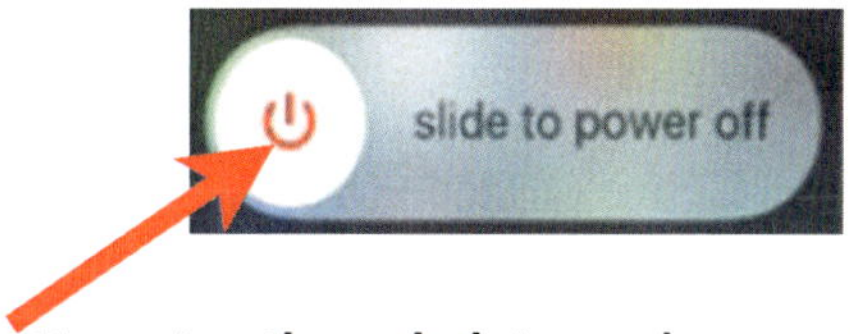

Slide this button to the right and your iPad will turn off.

Putting Your iPad Into Sleep Mode:

Press the Sleep/Wake Button briefly and your iPad will go into Sleep Mode.
Putting your iPad into Sleep Mode, when you're finished using it, helps to extend the battery life.

Note: The first time you turn on your iPad you'll be asked some questions in order to set up your iPad. We will go through a step-by-step set up in the next few chapters.

Setting Up Your iPad

Turn your iPad on. Press the home button to proceed.
(Refer to the diagram on page 3 for location of home button.)

Touch anywhere in the space next to the language you'd like to use on your iPad.

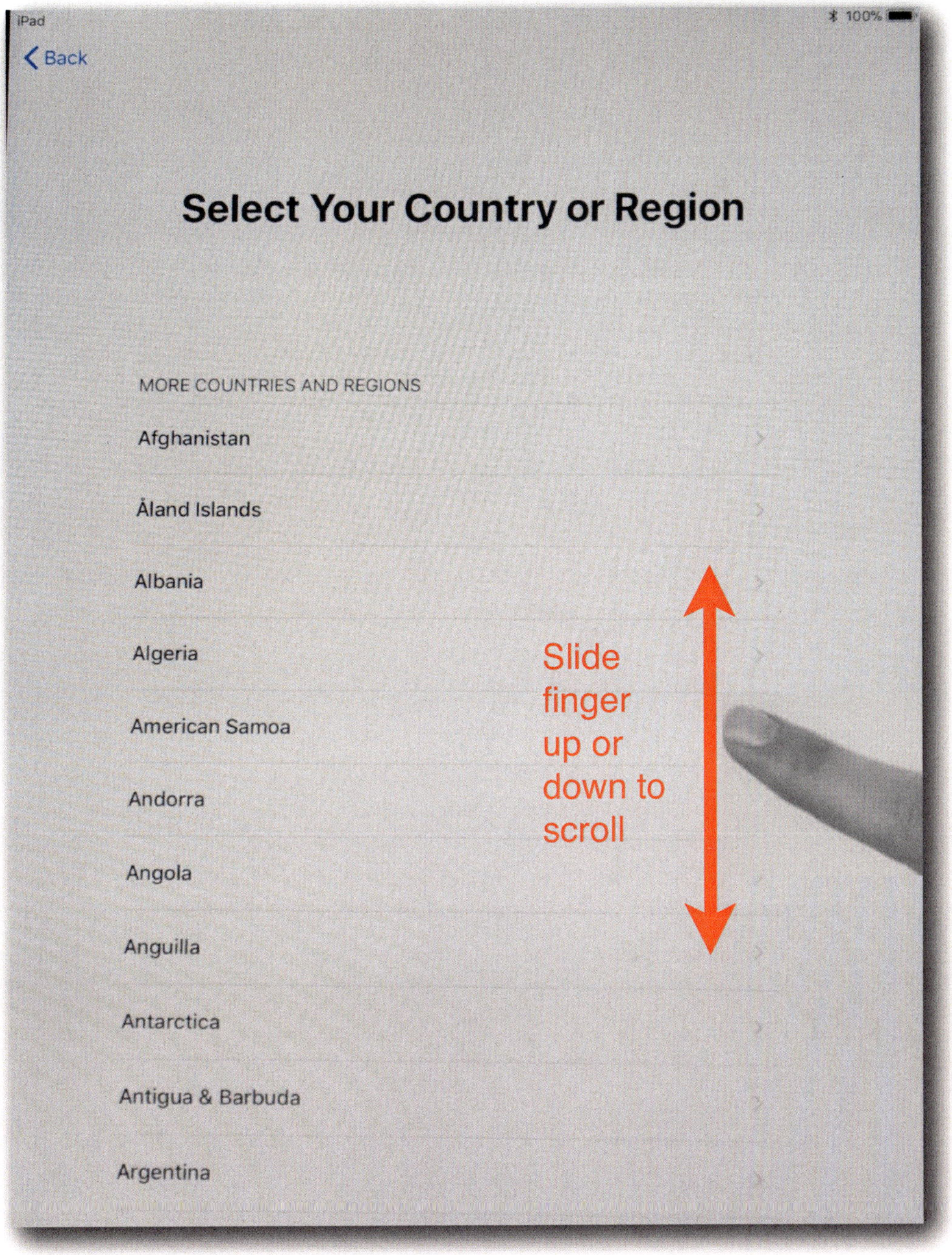

Scroll down the page until you come to your Country or Region. Then touch anywhere in the space next to your choice.

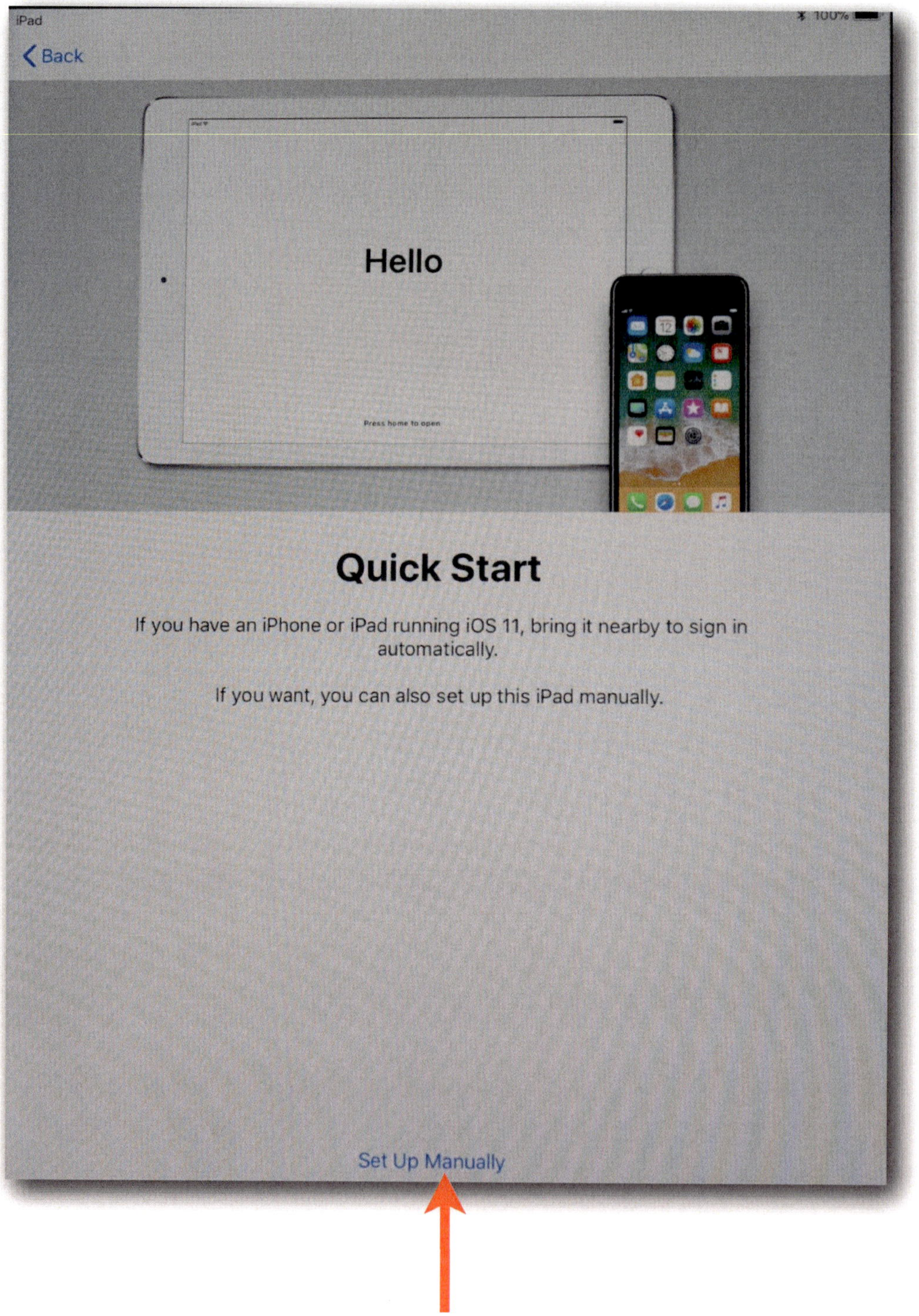
iPad
100%
Back
Hello
Press home to open
Quick Start
If you have an iPhone or iPad running iOS 11, bring it nearby to sign in automatically.
If you want, you can also set up this iPad manually.
Set Up Manually

Touch here.

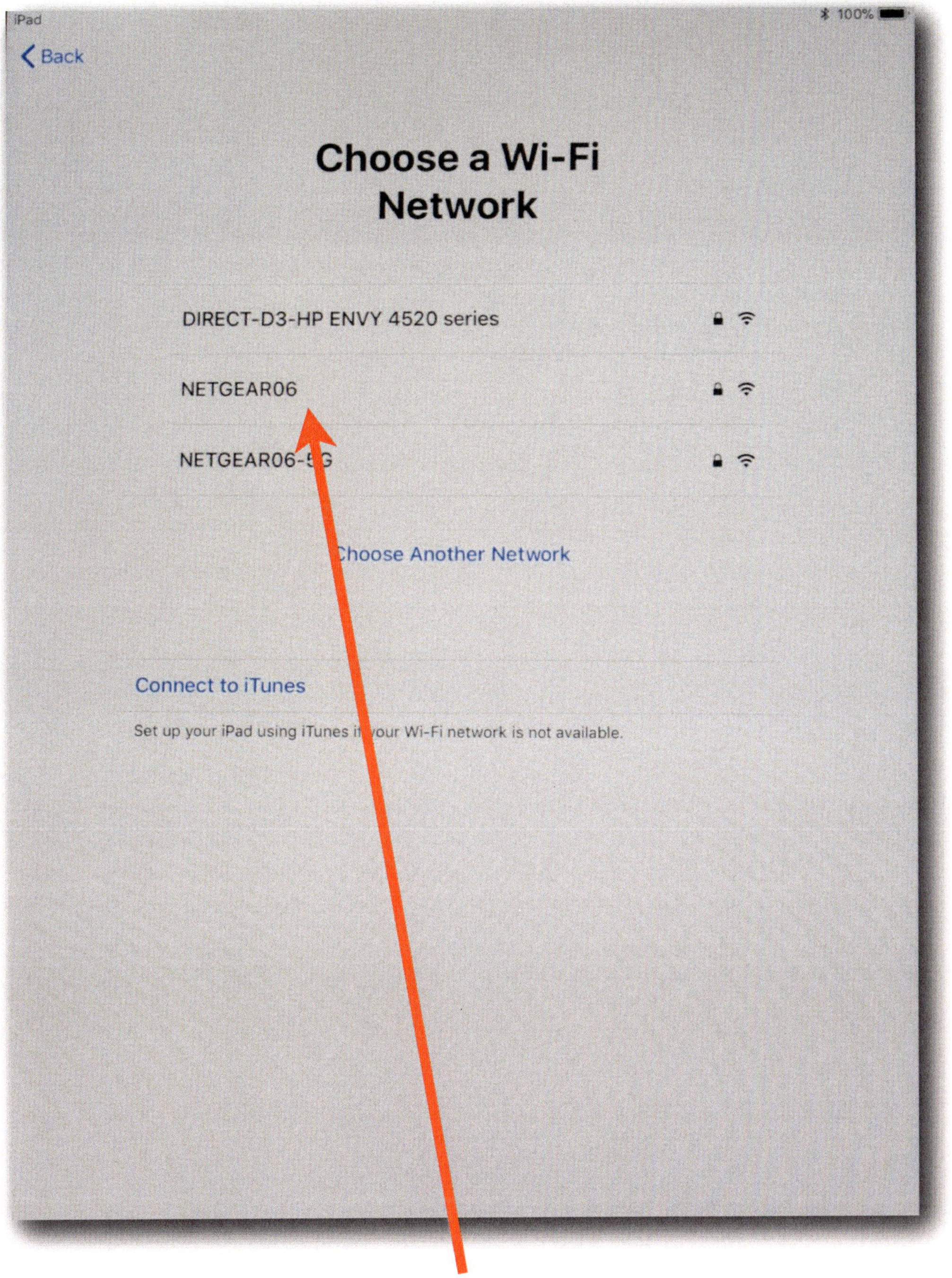

Touch your Wi-Fi Network. (This is mine. Yours will be different.)

Note: You will need a wifi connection in order to use many features on your iPad, such as email, text messaging and internet searches, to name a few. Your Wi-Fi router will have a password printed somewhere on it. This is the password you will need to enter on the next page.

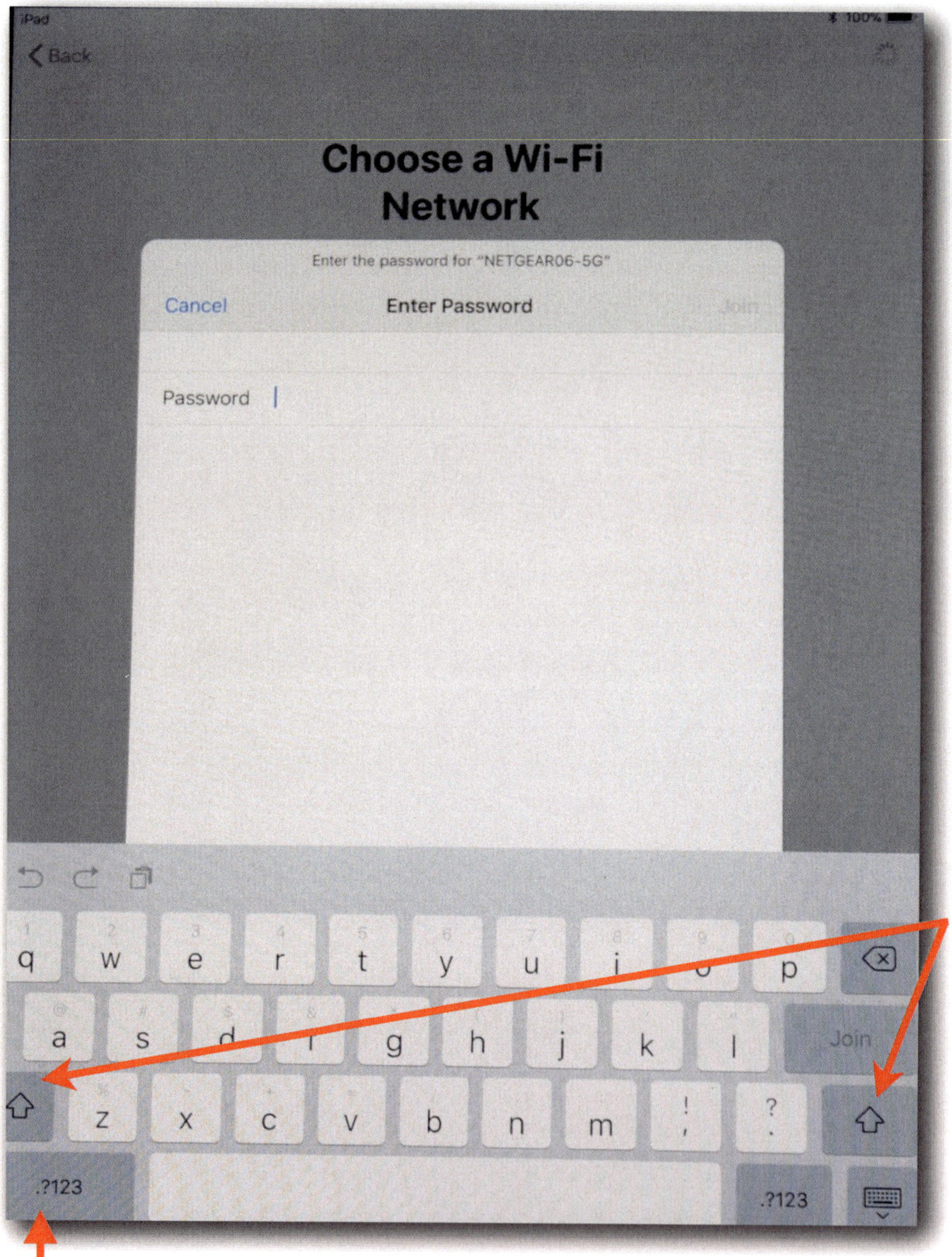

Touch either arrow key to access upper case letters.

Type in the password for your Wi-Fi Network on the keyboard that pops up on the bottom of your iPad screen.

Touch here
to access numbers and symbols.

Note: Your password will show up as dots as you type.

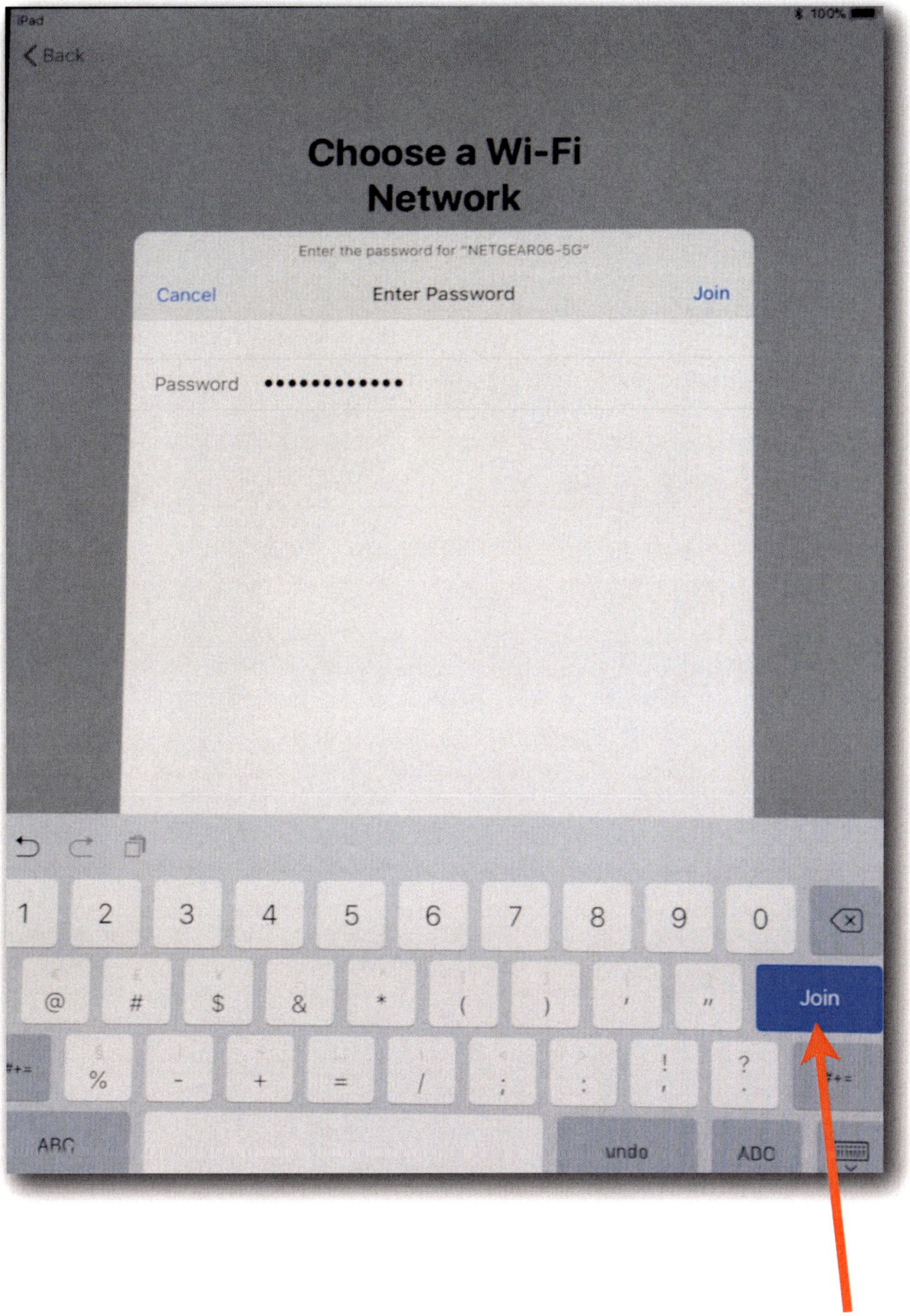

Once you're done typing in your Wi-Fi password, touch here.

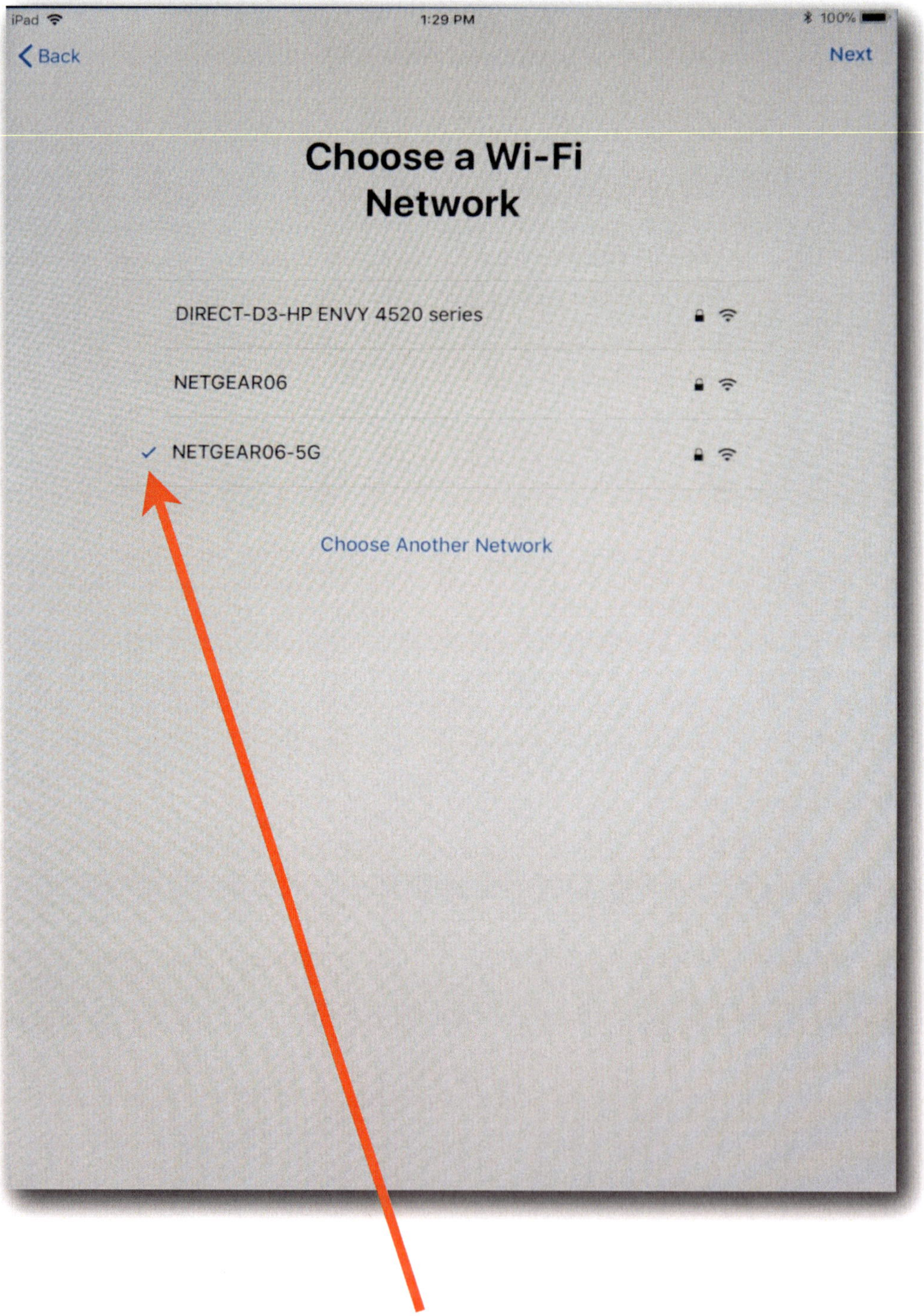

When your iPad has successfully connected to your Wi-Fi, a blue check mark will appear next to your network name.

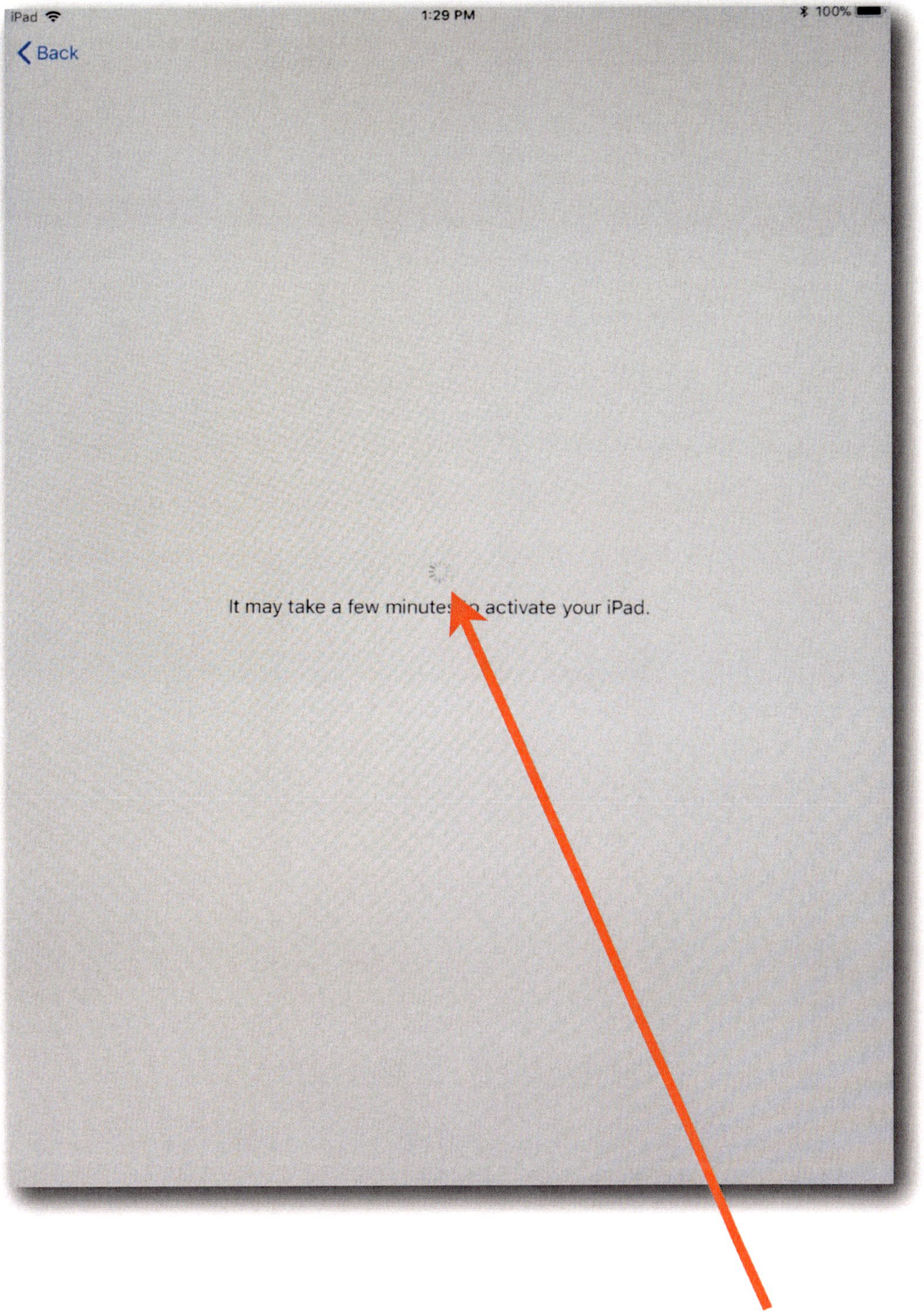
iPad
1:29 PM
100%
Back
It may take a few minutes to activate your iPad.

Be patient as your iPad finishes activating. The little spinning wheel appears when the iPad is working on something.

Skip this step for now by touching here.

iPad
1:30 PM
100%
Back
Touch ID
Use your fingerprint in place of your passcode or Apple ID password for purchases.
Continue
Set Up Touch ID Later
Are you sure you don't want to use Touch ID?
Touch ID secures your iPad and can be used in place of your passcode or Apple ID password for purchases.
Don't Use
Use Touch ID

Touch here for now.

Touch here for options.

You may wish to set up a passcode, which you will have to type in every time you want to use your iPad. I recommend skipping this step for now. You can always set it up later.

If you choose not to use a passcode, touch here and skip to page 23.

If you choose to set up a passcode, turn to the next page.

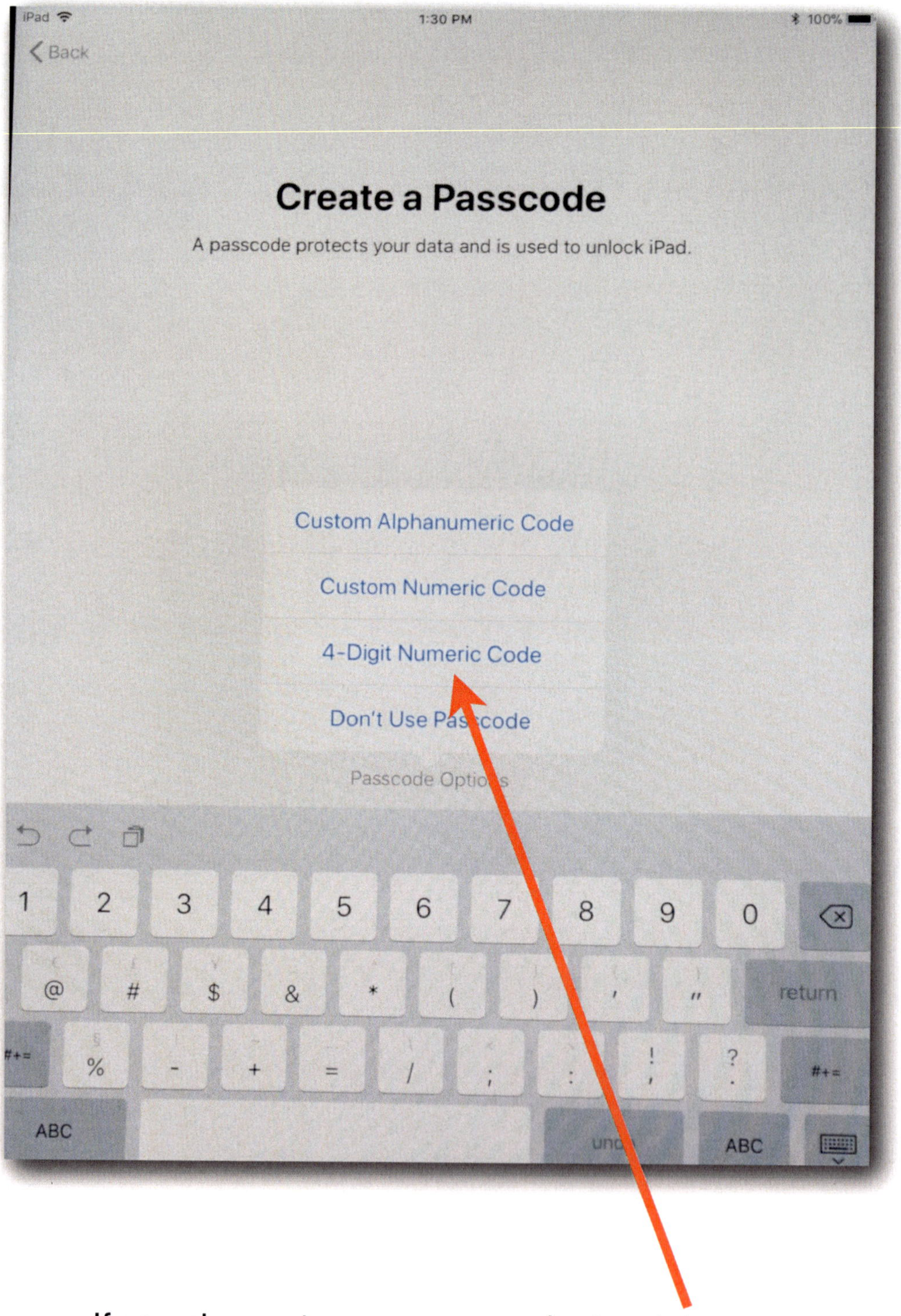

If you choose to use a passcode, touch here.

Using the keyboard, type in your chosen passcode.

Note: Be sure your passcode is easy to remember **and** write it down and keep it somewhere safe.

Re-enter your passcode on this page.

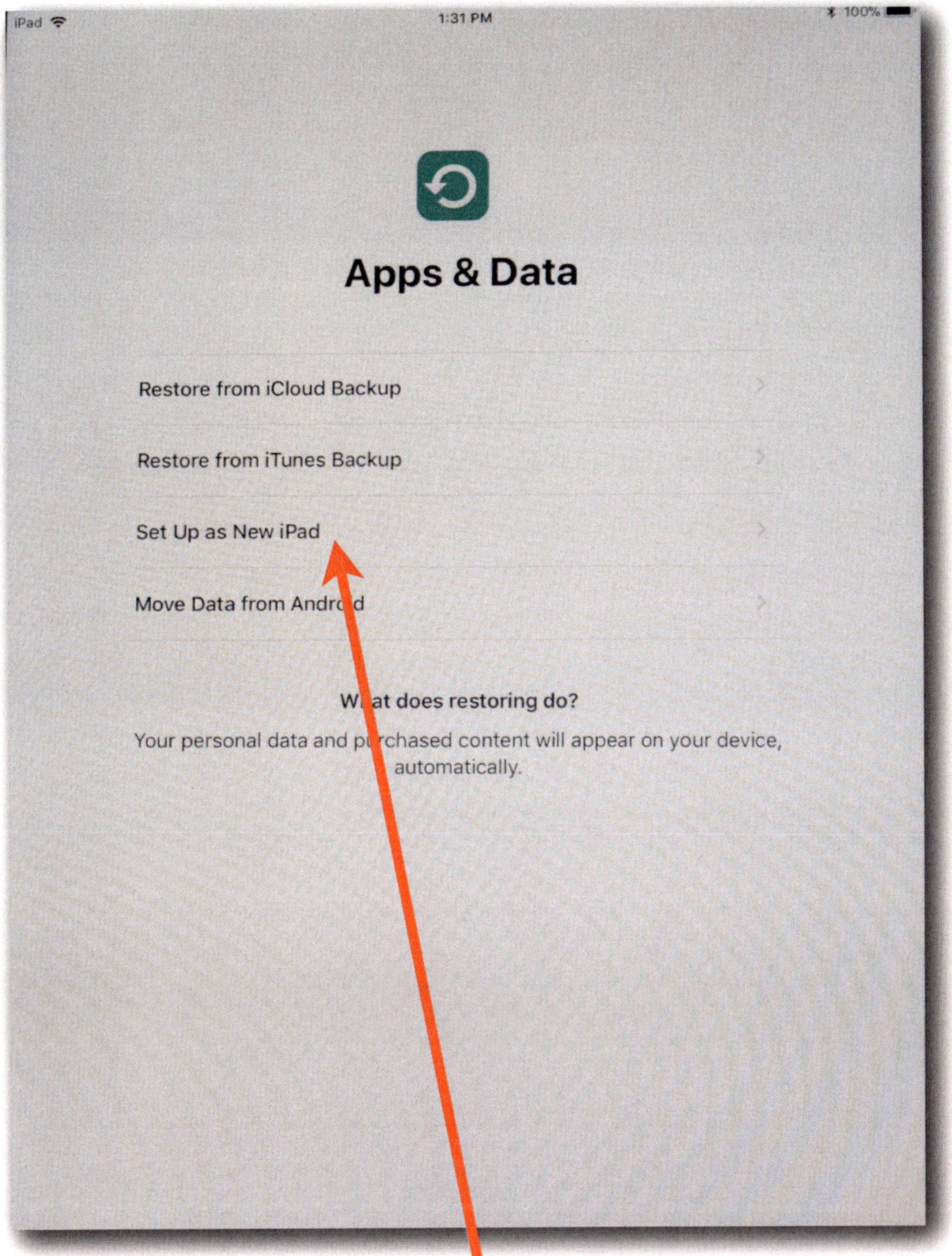
iPad
1:31 PM
100%
Apps & Data
Restore from iCloud Backup
Restore from iTunes Backup
Set Up as New iPad
Move Data from Android
What does restoring do?
Your personal data and purchased content will appear on your device, automatically.

Touch here.

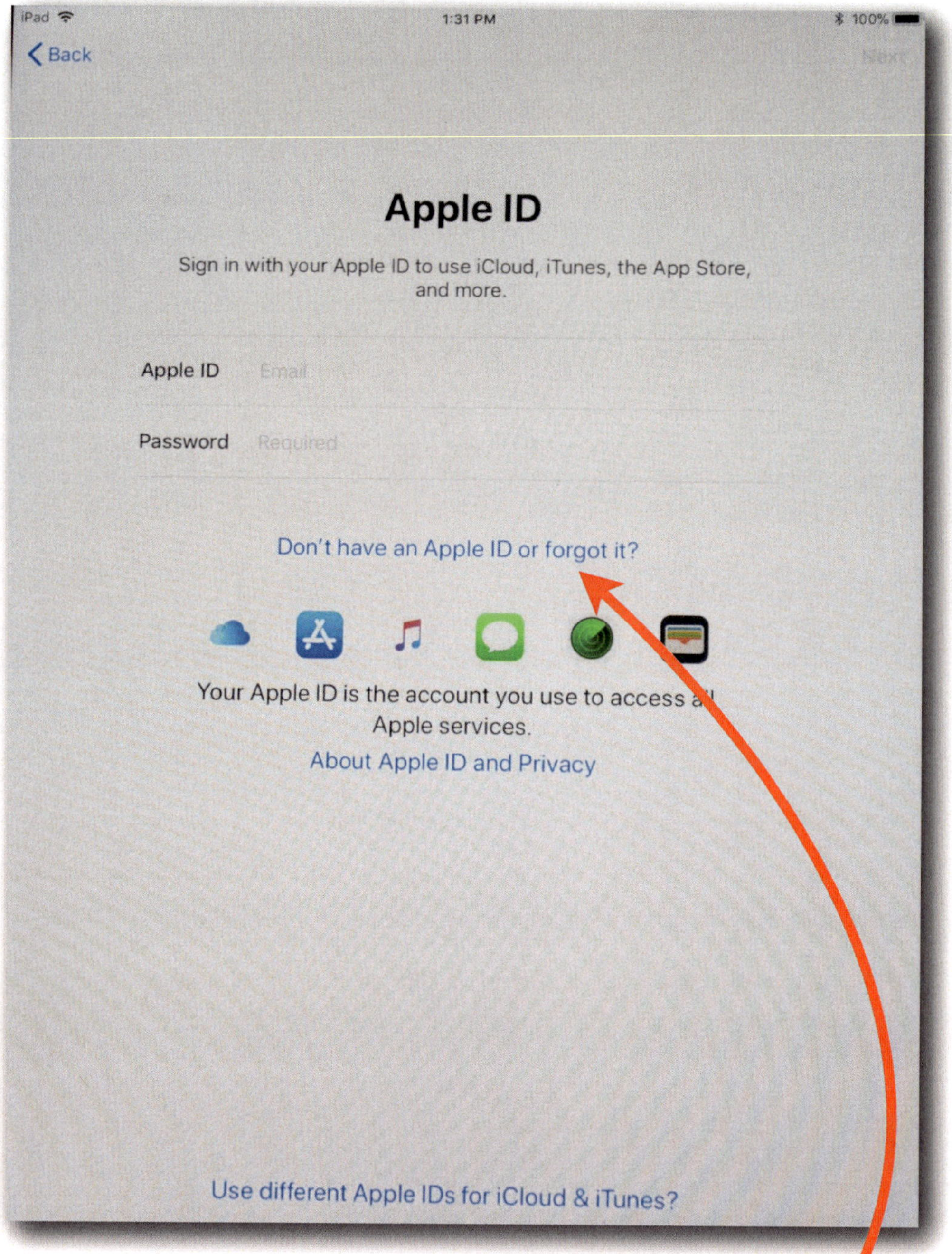
iPad
1:31 PM
100%
Back
Apple ID
Sign in with your Apple ID to use iCloud, iTunes, the App Store, and more.
Apple ID
Email
Password
Required
Don't have an Apple ID or forgot it?
Your Apple ID is the account you use to access all Apple services.
About Apple ID and Privacy
Use different Apple IDs for iCloud & iTunes?

If you already have an Apple ID, enter your information on this page.

If you do not have an Apple ID, touch here.

Note: Apple ID is an email address registered with Apple, for sending and receiving text messages and adding apps to your device.

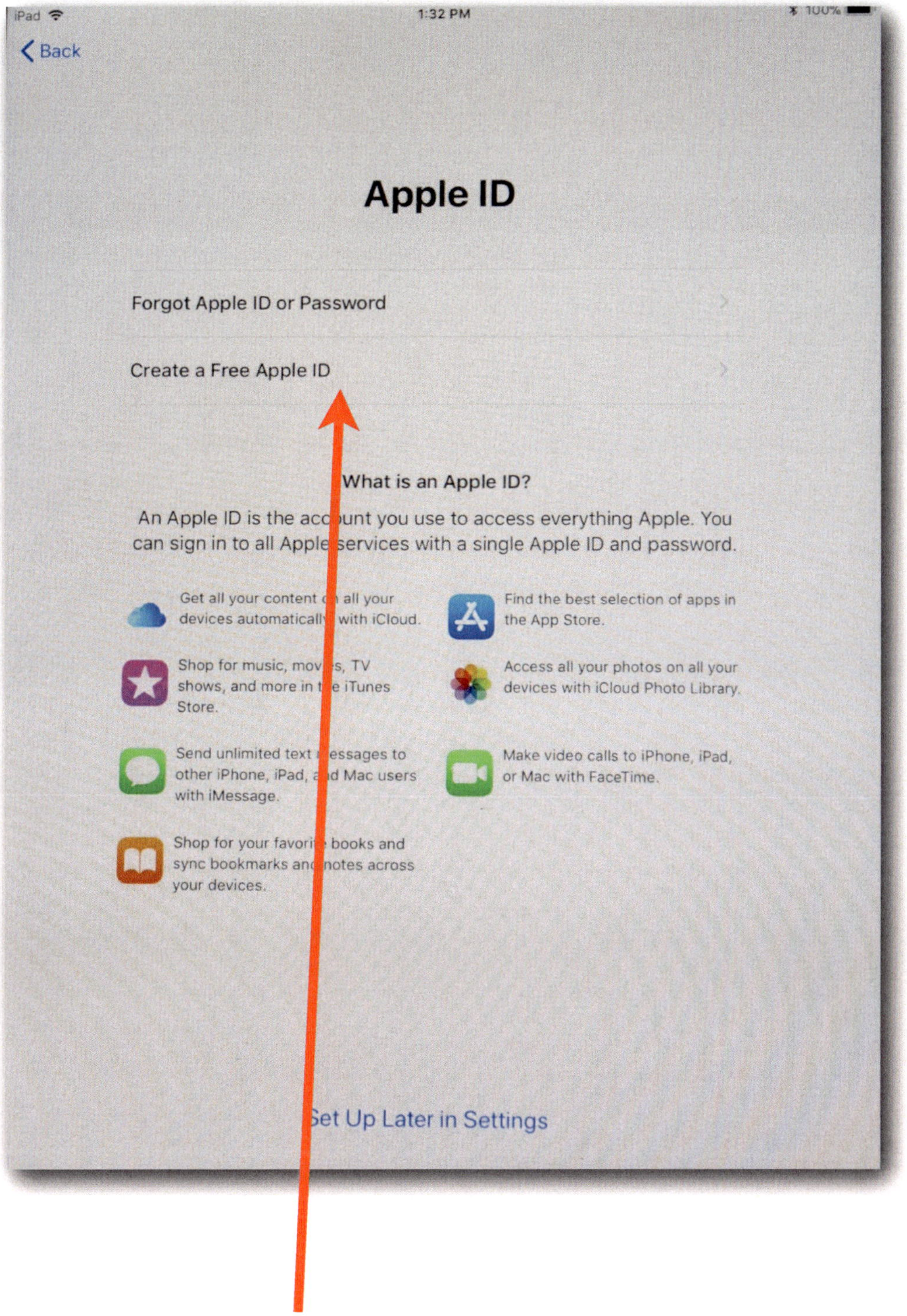
iPad
1:32 PM
100%
Back
Apple ID
Forgot Apple ID or Password
Create a Free Apple ID
What is an Apple ID?
An Apple ID is the account you use to access everything Apple. You can sign in to all Apple services with a single Apple ID and password.
Find the best selection of apps in the App Store.
Access all your photos on all your devices with iCloud Photo Library.
Make video calls to iPhone, iPad, or Mac with FaceTime.
Set Up Later in Settings

Touch here to set up a new Apple ID.

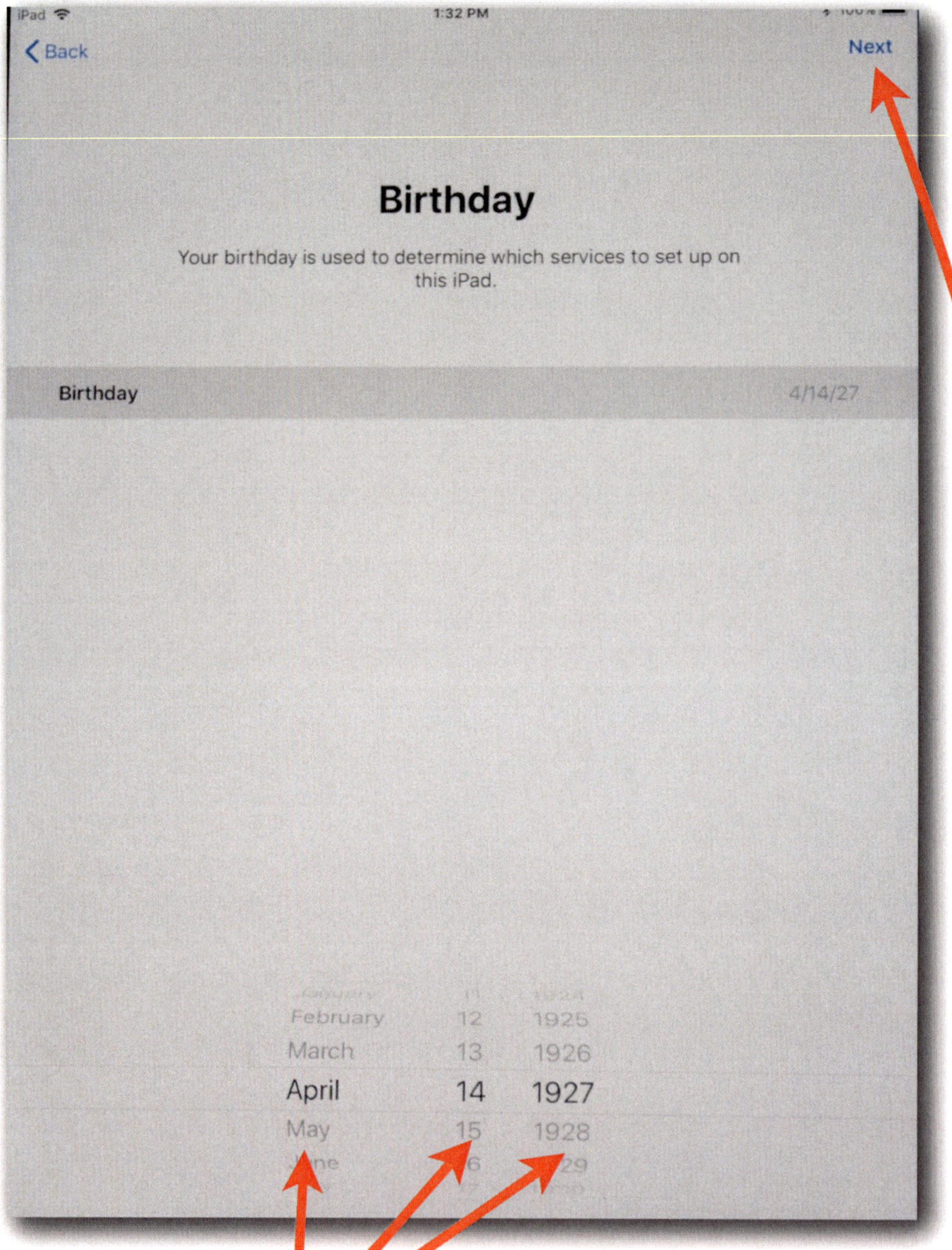

Scroll up or down here to select the month, day and year of your birthday.

When finished, touch here.

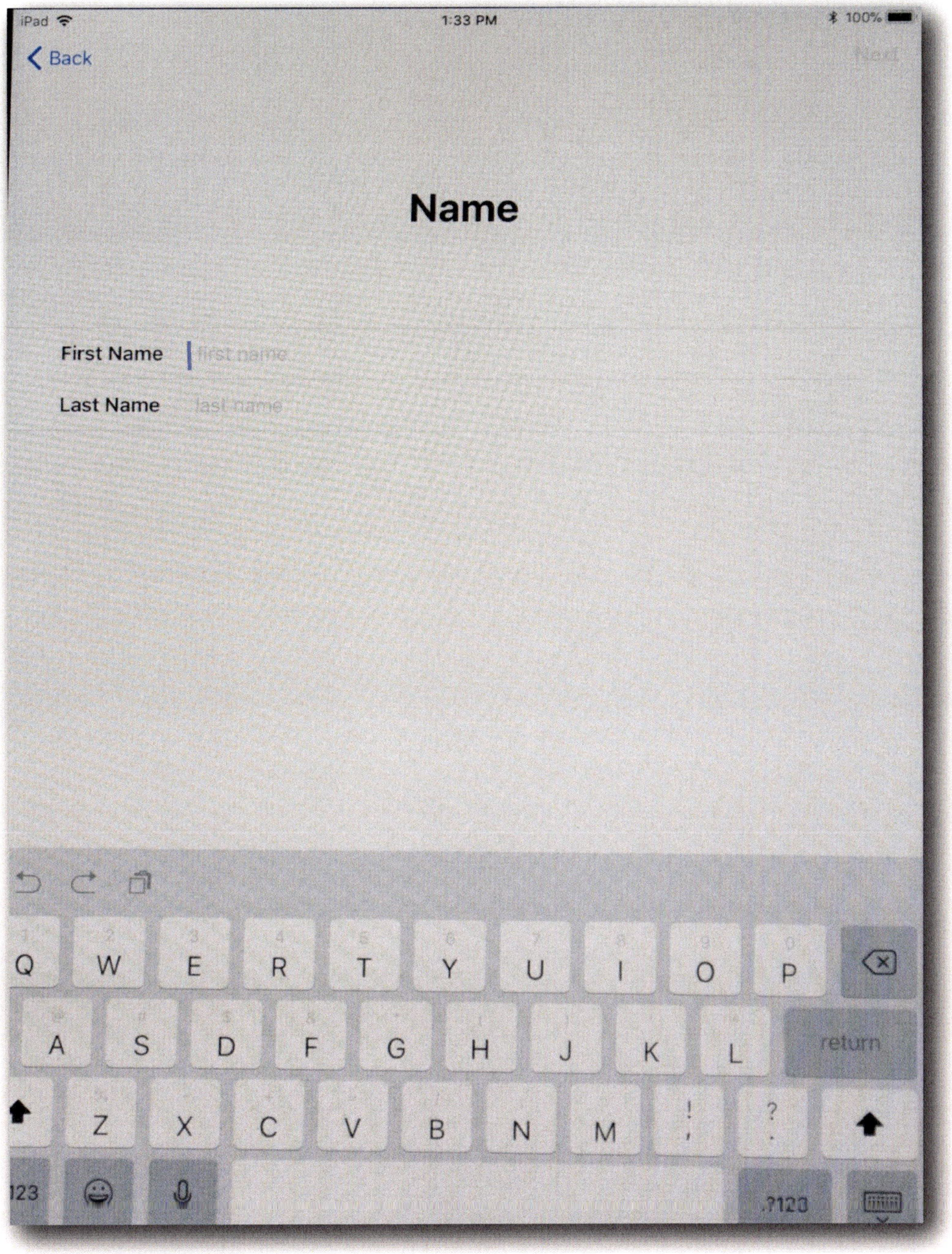

Type your first name.

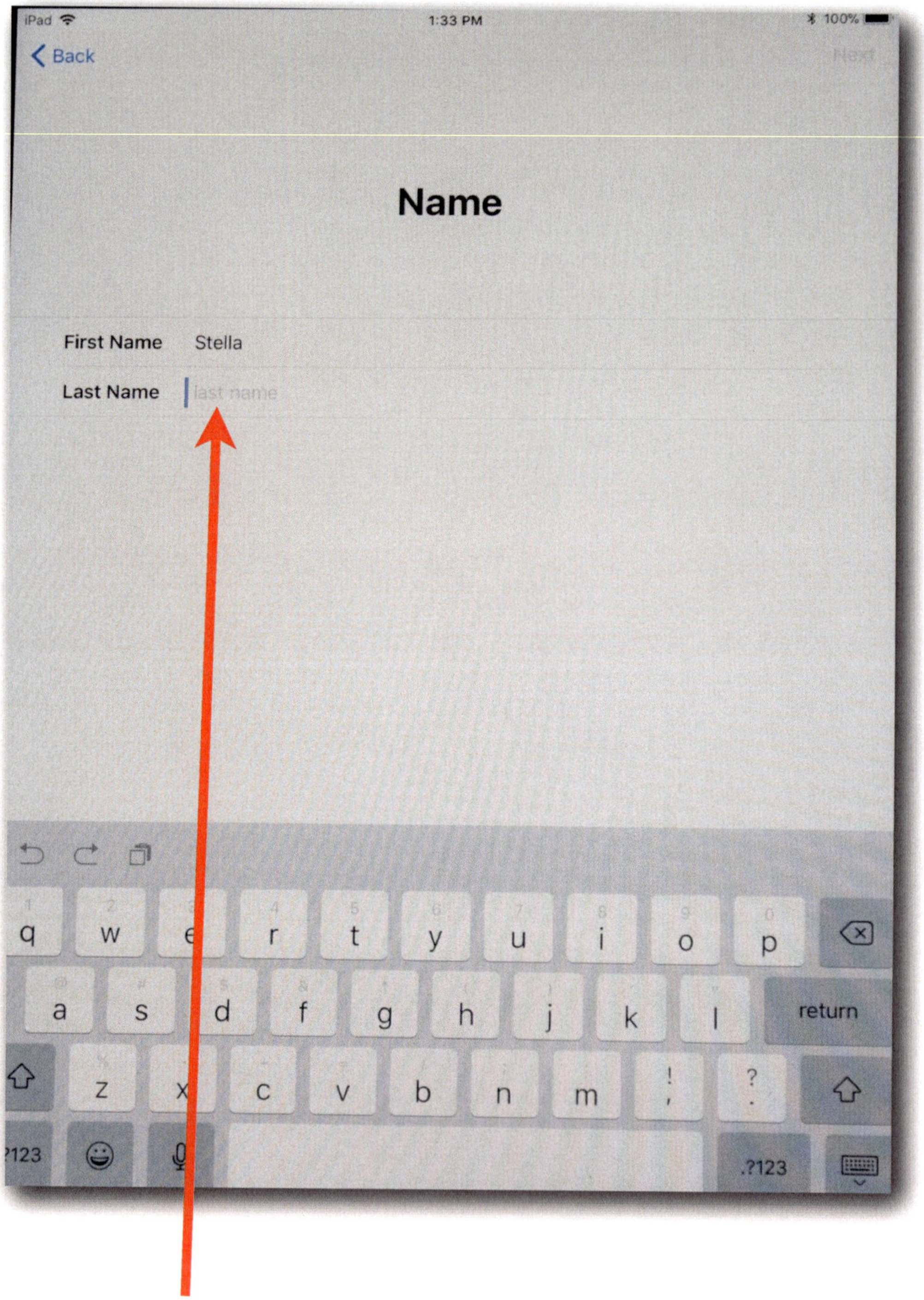

Touch here. When the blue cursor blinks in the space you touched, type your last name.

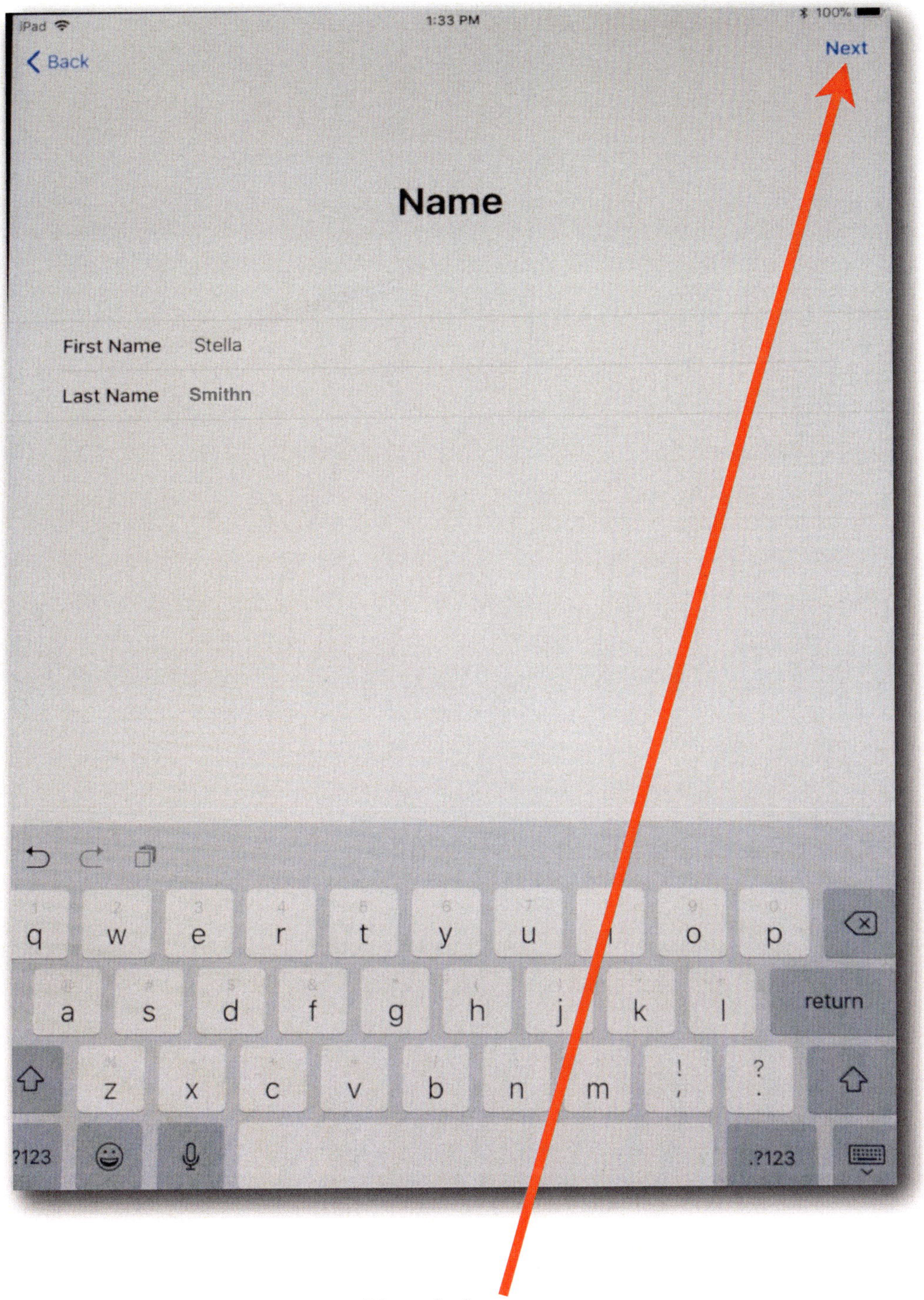
iPad
1:33 PM
100%
Back
Next
Name
First Name
Stella
Last Name
Smithn
return
.?123

Touch here.

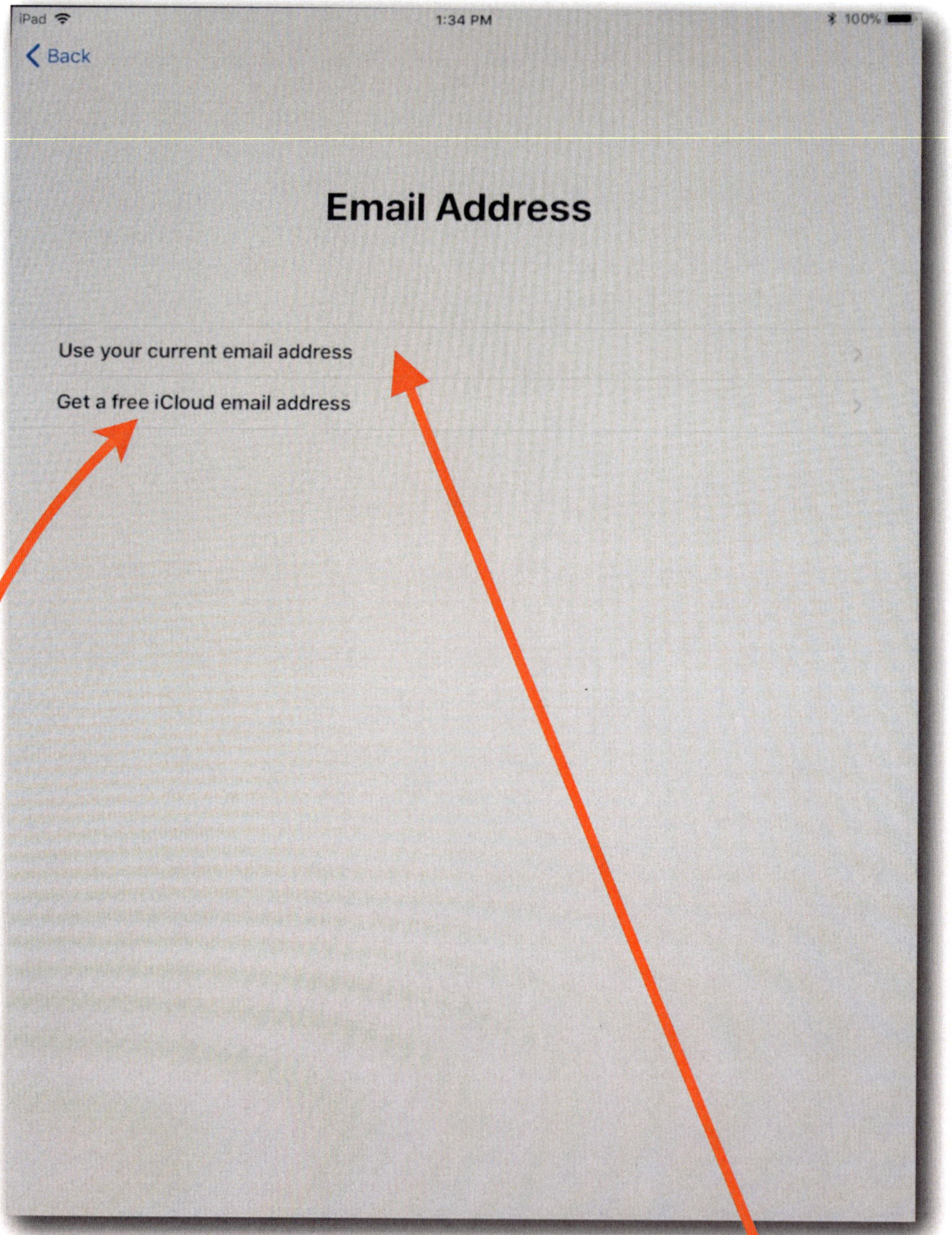

If you already have an email address, touch here.

If you don't have an email address and would like to to set up an iCloud email address, touch here.

Note: iCloud Mail is a free email service from Apple.

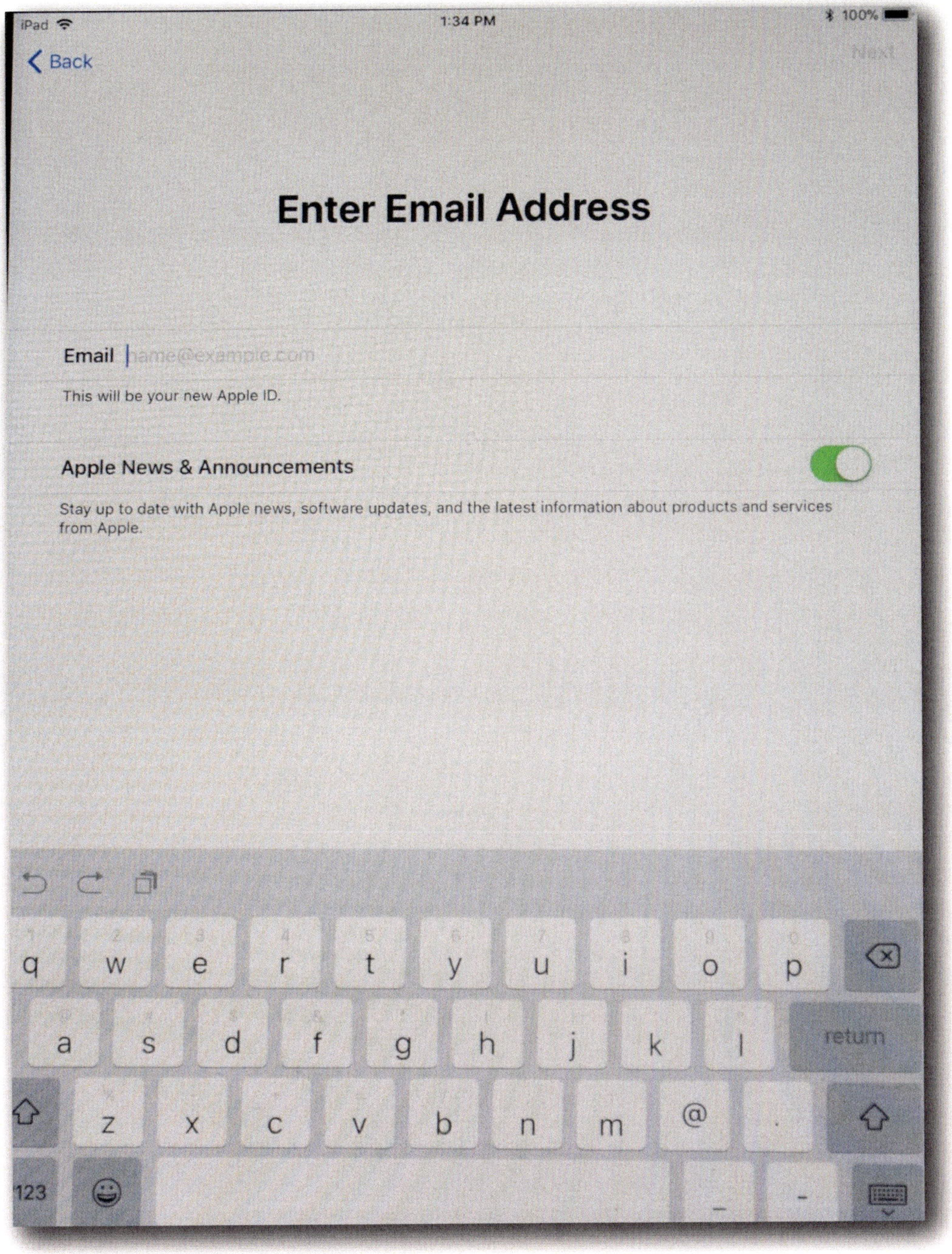
iPad
1:34 PM
100%
Back
Next
Enter Email Address
Email name@example.com
This will be your new Apple ID.
Apple News & Announcements
Stay up to date with Apple news, software updates, and the latest information about products and services from Apple.
return
123

If you've chosen to use your current email address, type your current email address.

or…(next page)

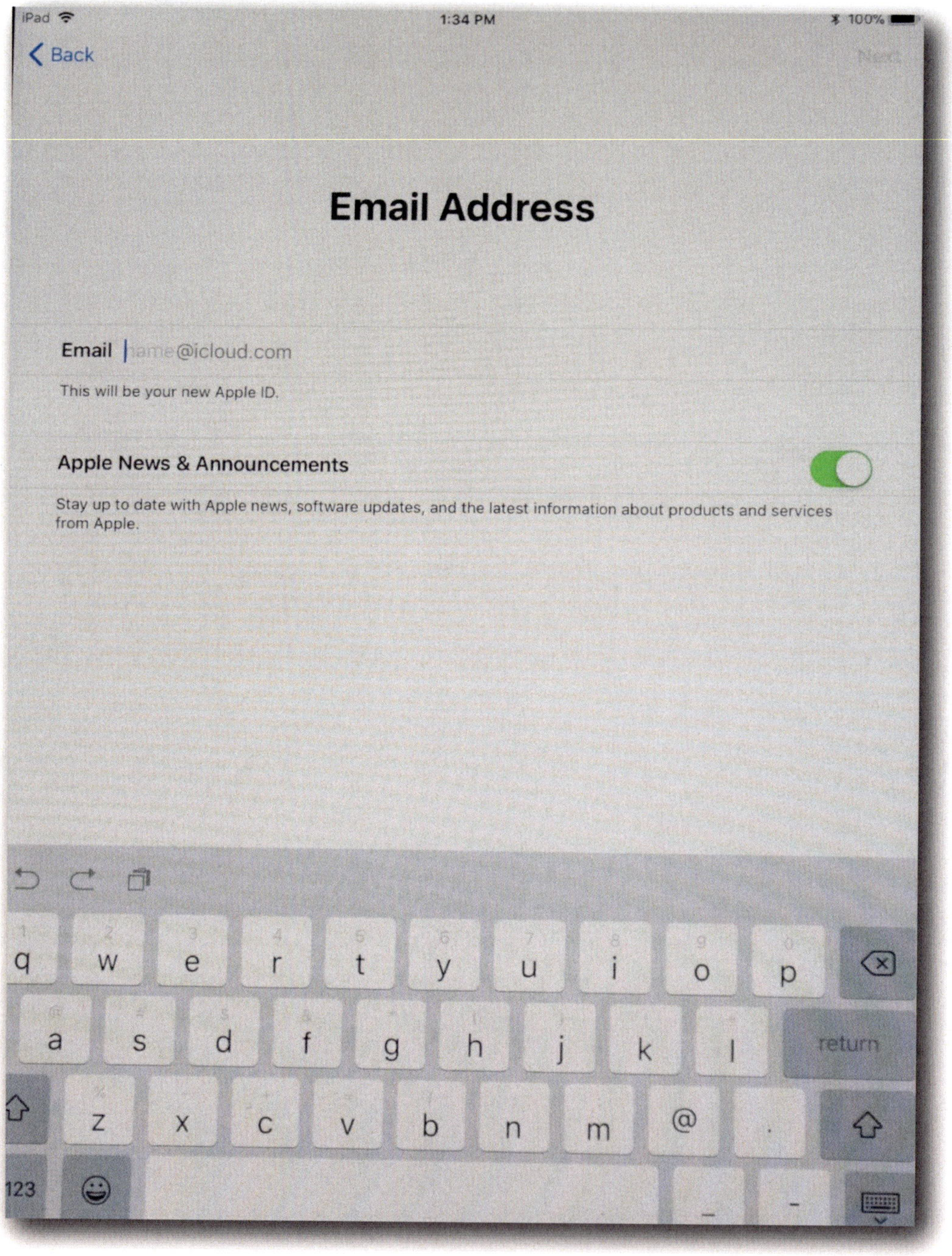
iPad
1:34 PM
100%
Back
Next
Email Address
Email name@icloud.com
This will be your new Apple ID.
Apple News & Announcements
Stay up to date with Apple news, software updates, and the latest information about products and services from Apple.
return
123

...if you've chosen to create a new iCloud email address, create your new email by typing it in on this page.

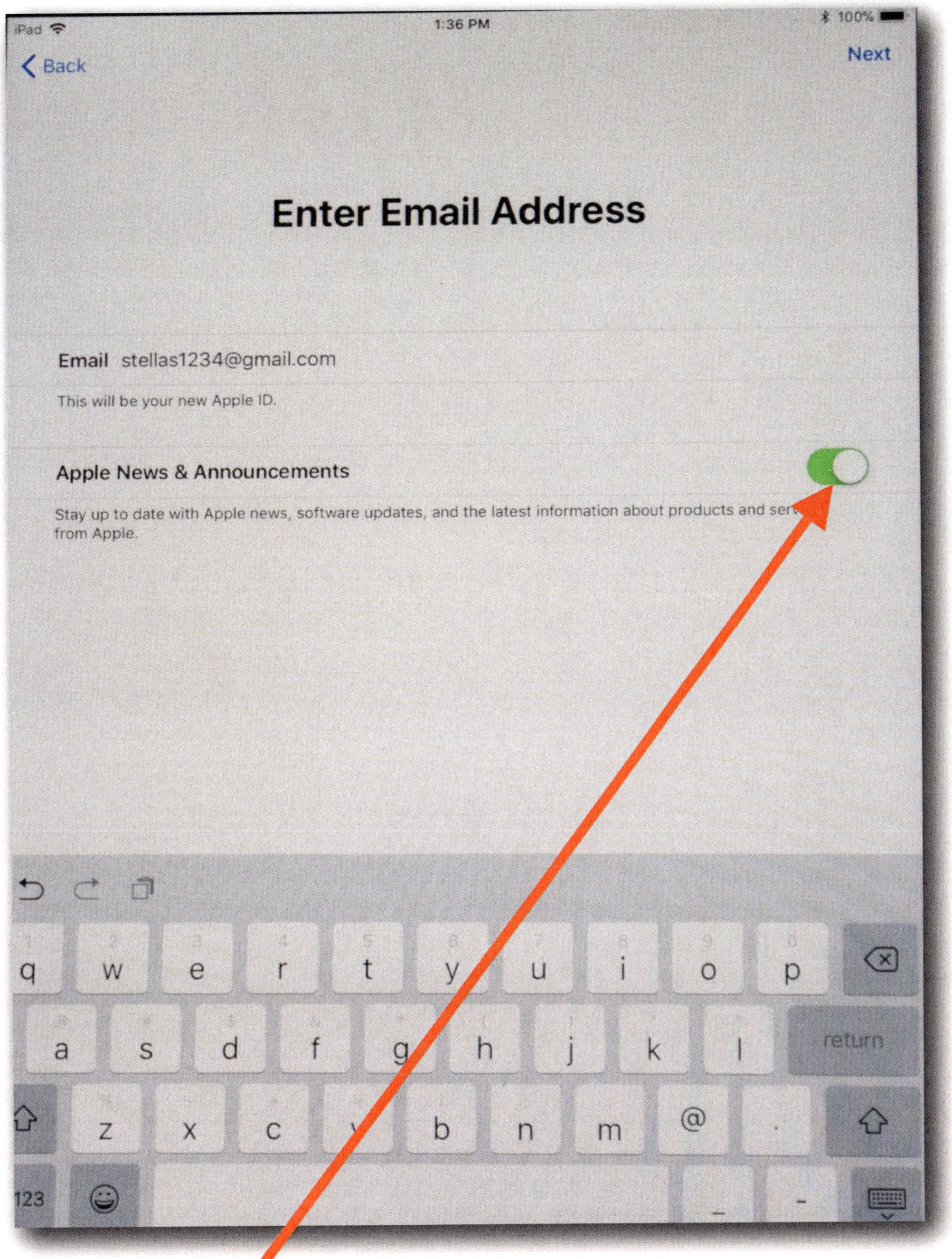
iPad
1:36 PM
100%
Back
Next
Enter Email Address
Email stellas1234@gmail.com
This will be your new Apple ID.
Apple News & Announcements
Stay up to date with Apple news, software updates, and the latest information about products and ser
from Apple.
return
123

Touch here to turn off Apple News and Announcements for now.

Note: This book is for beginners who want to learn just the absolute basics for using an iPad. Options such as this one can be turned on later, once the basics have been mastered.

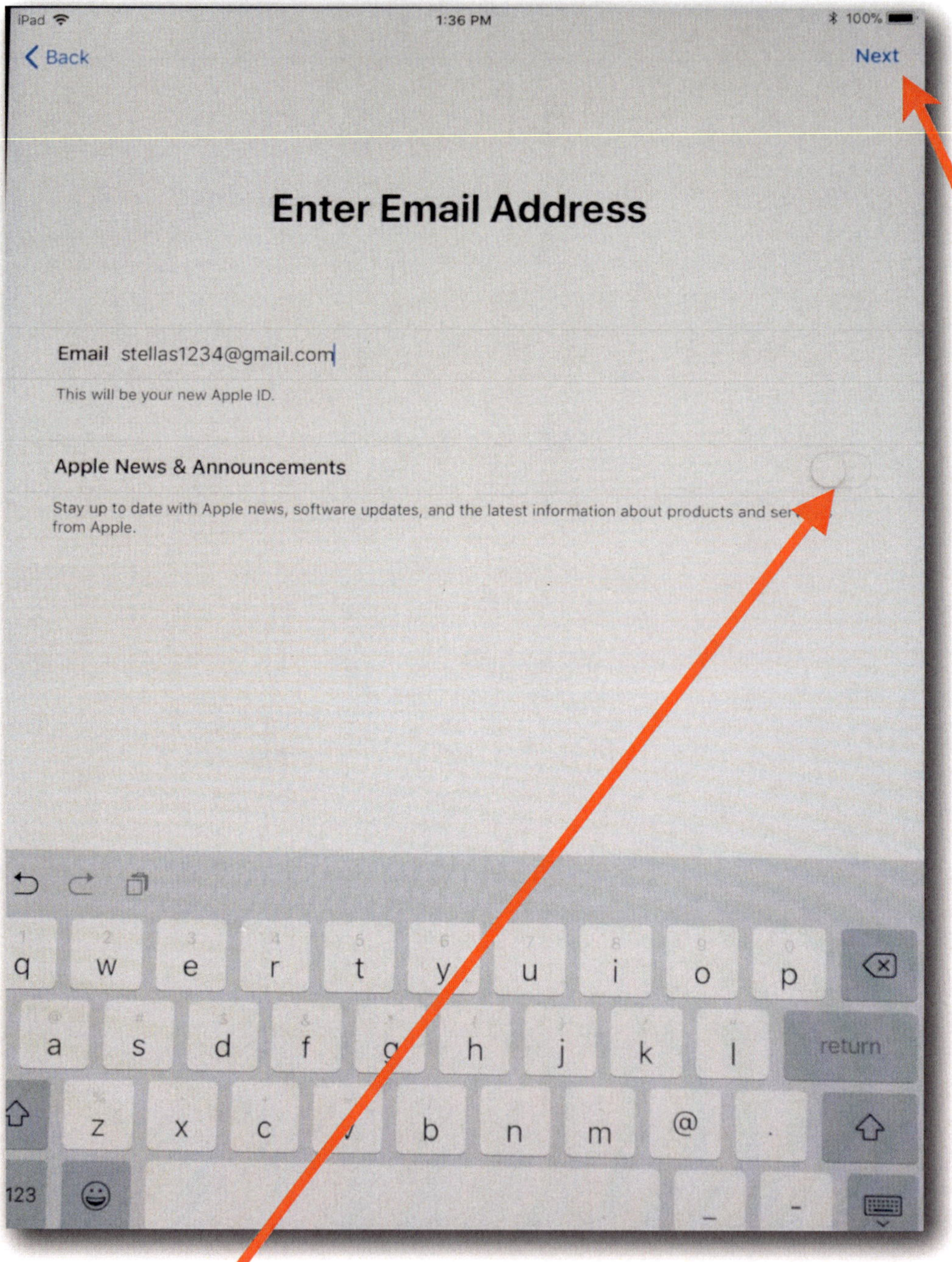

When this button is no longer green, it has been successfully turned off.

Touch here to proceed.

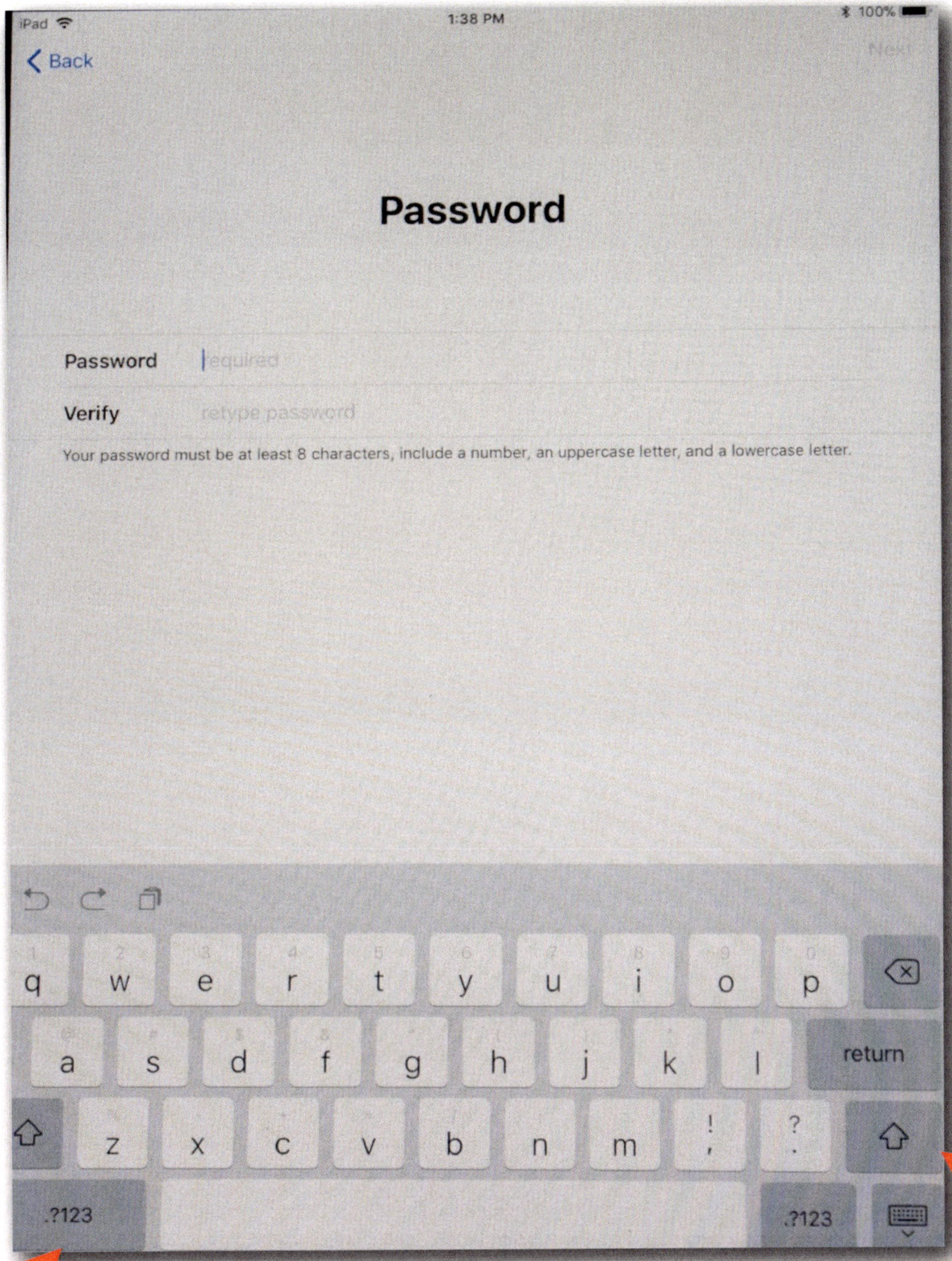
iPad
1:38 PM
100%
Back
Next
Password
Password
required
Verify
retype password
Your password must be at least 8 characters, include a number, an uppercase letter, and a lowercase letter.
return
.?123
.?123

Create an easy-to-remember password for your new Apple ID account and type it in on this page.

Note: It must be at least 8 characters and must include at least one number, one upper case letter and one lower case letter. To access upper case letters, touch one of the arrow keys. Touch here for numbers.

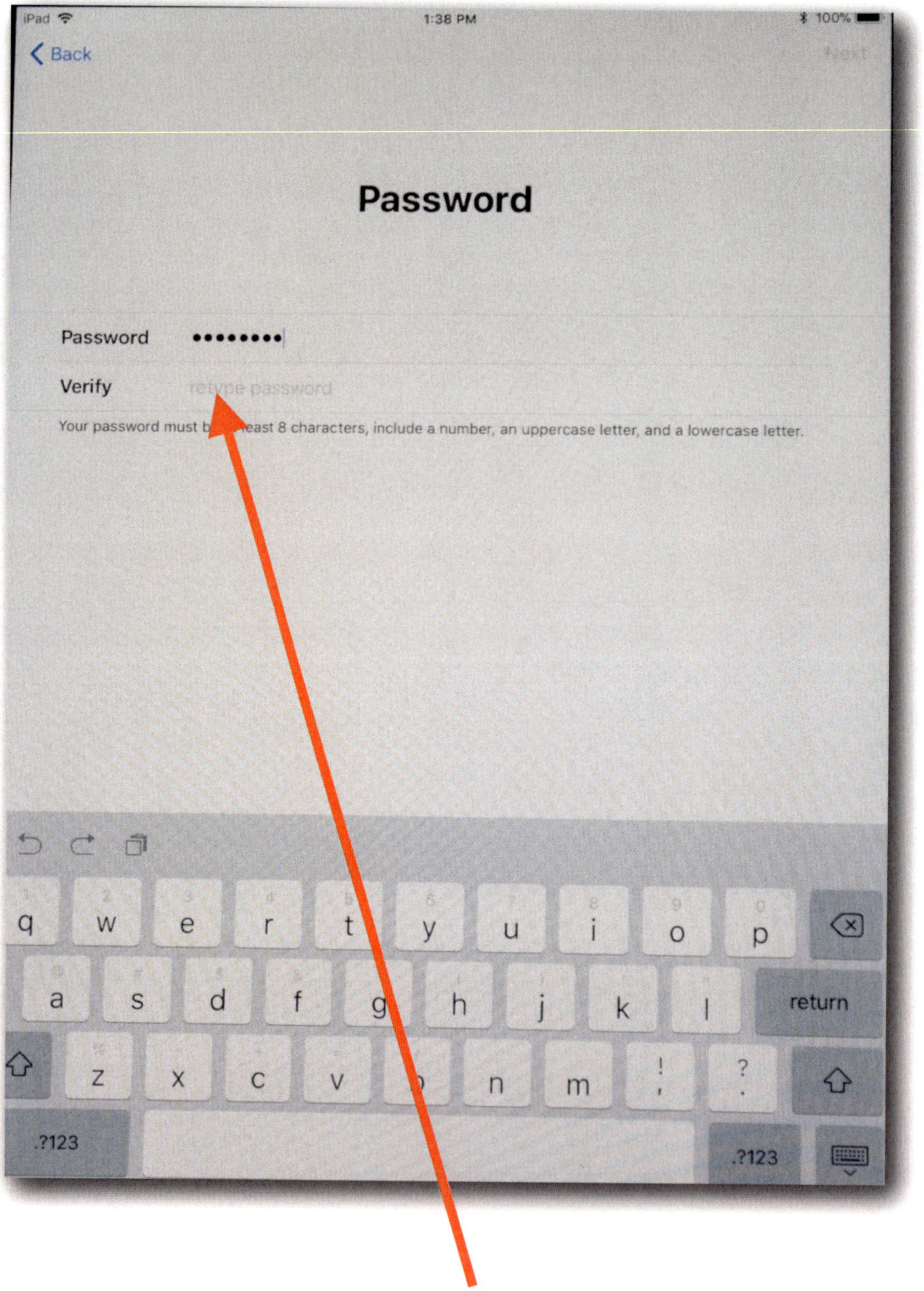
iPad
1:38 PM
100%
Back
Next
Password
Password
Verify
retype password
return
.?123
.?123

Touch here.

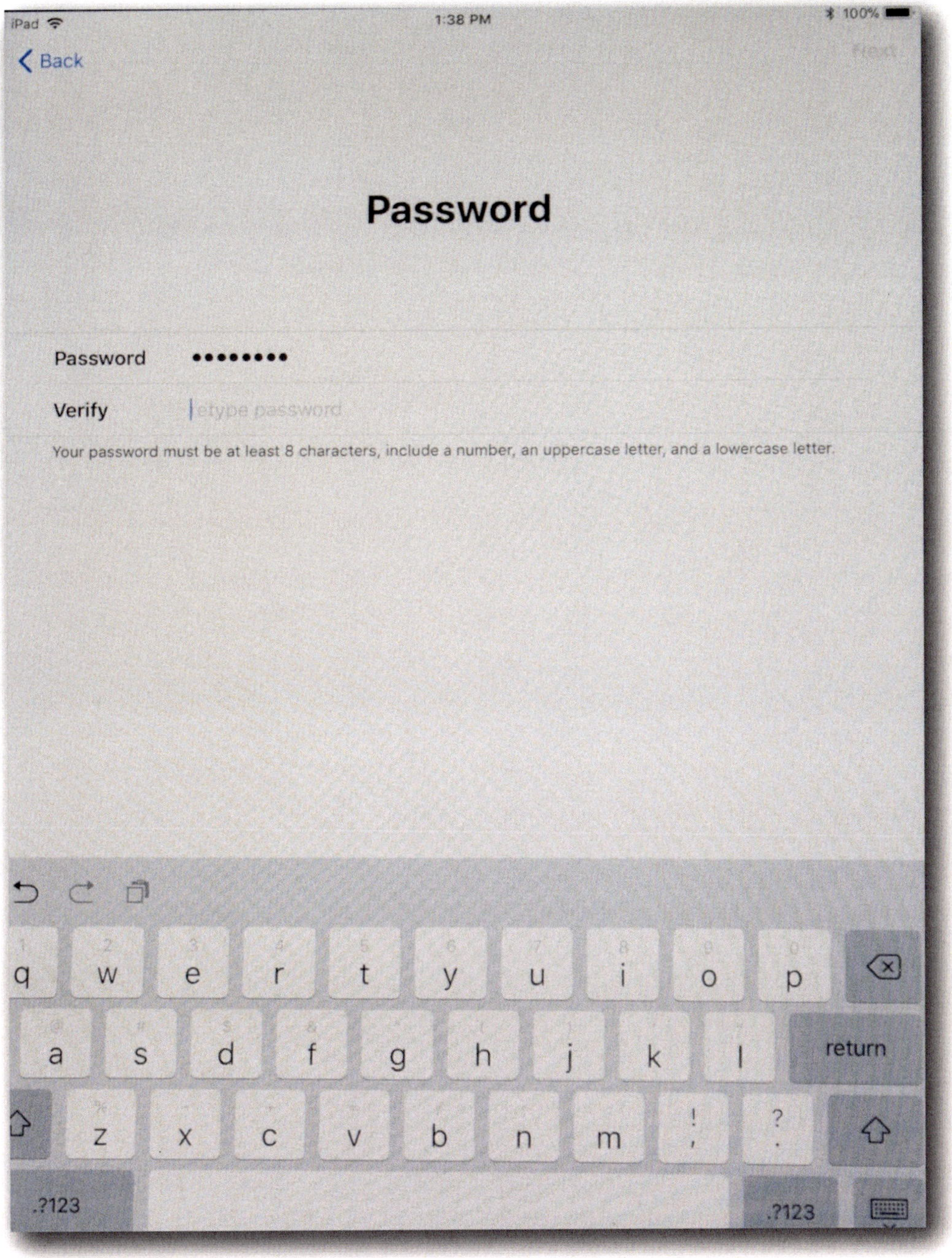
iPad
1:38 PM
100%
Back
Password
Password
Verify
Your password must be at least 8 characters, include a number, an uppercase letter, and a lowercase letter.
return
.?123
.?123

Re-type your password.

iPad
1:39 PM
100%
Back
Next
Password
Password
Verify
Your password must be at least 8 characters, include a number, an uppercase letter, and a lowercase letter.
return
undo
ABC

Touch here.

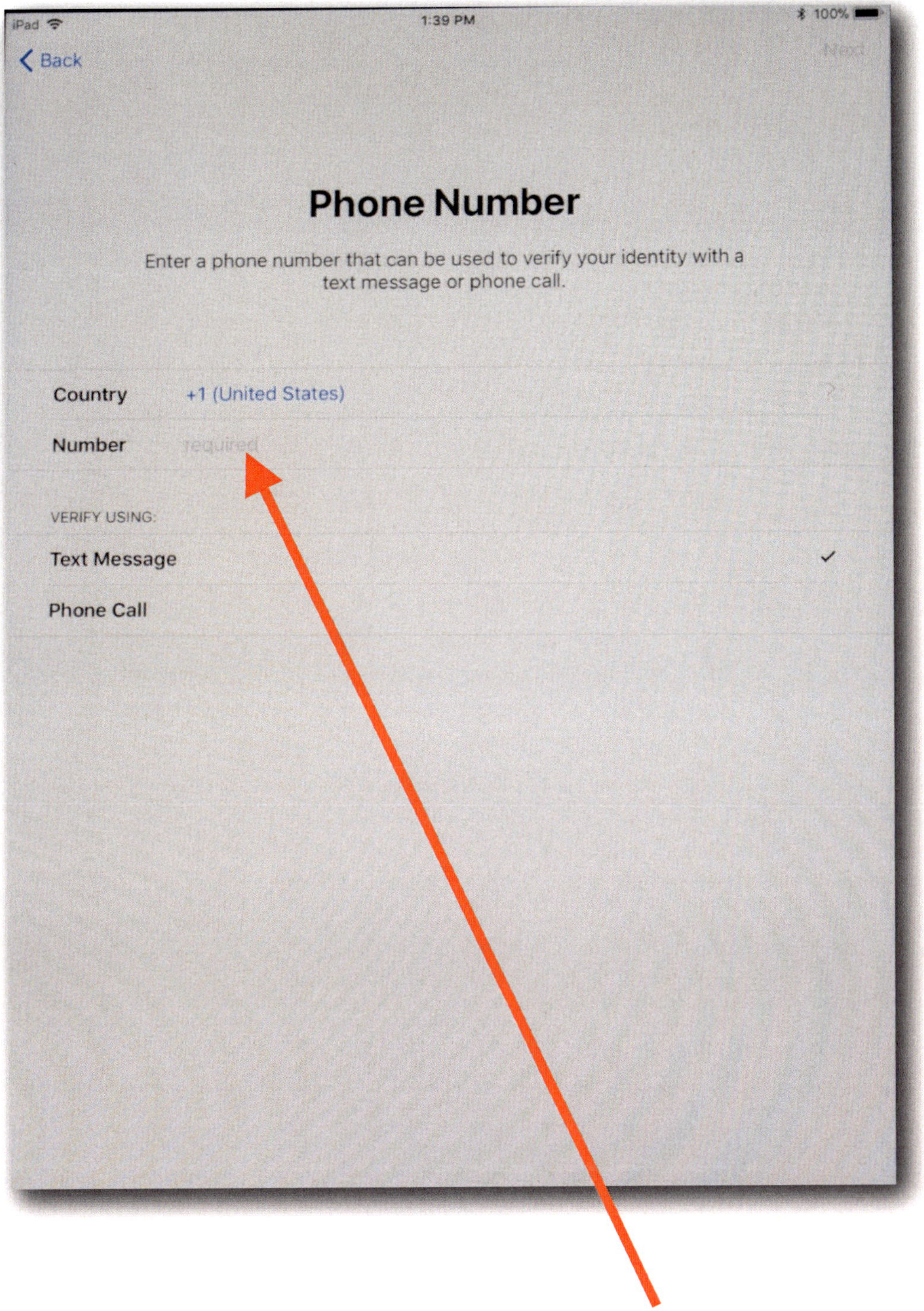

Enter your phone number here.

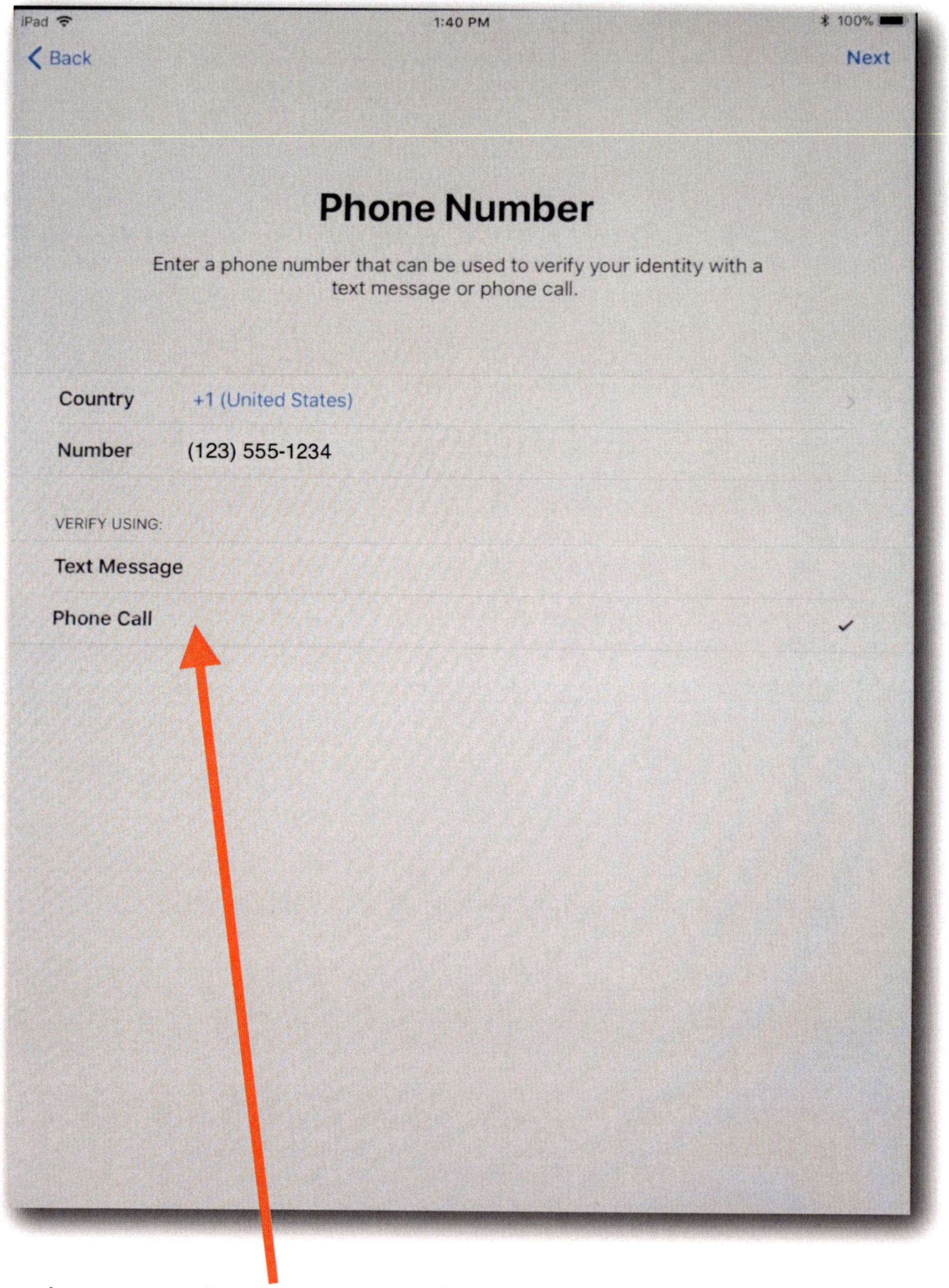

If the phone number you entered is a cell phone, touch your choice for communication here, either Text Message or Phone Call.

If the phone number you entered is a landline, touch Phone Call because typically you cannot receive text messages on a landline.

A verification code has been sent to your phone. Type in the verification code. Then touch here.

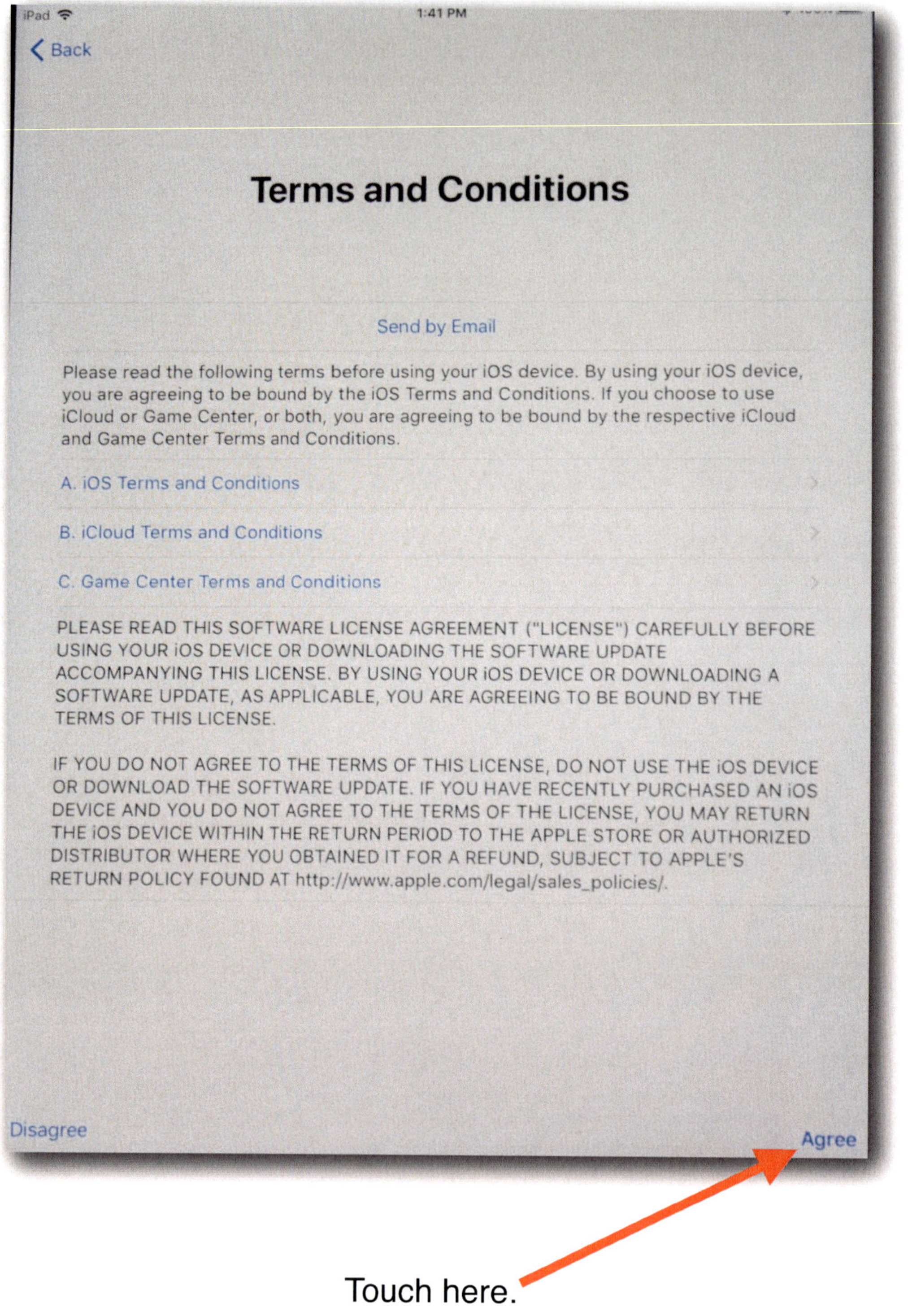
iPad
1:41 PM
Back
Terms and Conditions
Send by Email
Please read the following terms before using your iOS device. By using your iOS device, you are agreeing to be bound by the iOS Terms and Conditions. If you choose to use iCloud or Game Center, or both, you are agreeing to be bound by the respective iCloud and Game Center Terms and Conditions.
A. iOS Terms and Conditions
B. iCloud Terms and Conditions
C. Game Center Terms and Conditions
PLEASE READ THIS SOFTWARE LICENSE AGREEMENT ("LICENSE") CAREFULLY BEFORE USING YOUR iOS DEVICE OR DOWNLOADING THE SOFTWARE UPDATE ACCOMPANYING THIS LICENSE. BY USING YOUR iOS DEVICE OR DOWNLOADING A SOFTWARE UPDATE, AS APPLICABLE, YOU ARE AGREEING TO BE BOUND BY THE TERMS OF THIS LICENSE.
IF YOU DO NOT AGREE TO THE TERMS OF THIS LICENSE, DO NOT USE THE iOS DEVICE OR DOWNLOAD THE SOFTWARE UPDATE. IF YOU HAVE RECENTLY PURCHASED AN iOS DEVICE AND YOU DO NOT AGREE TO THE TERMS OF THE LICENSE, YOU MAY RETURN THE iOS DEVICE WITHIN THE RETURN PERIOD TO THE APPLE STORE OR AUTHORIZED DISTRIBUTOR WHERE YOU OBTAINED IT FOR A REFUND, SUBJECT TO APPLE'S RETURN POLICY FOUND AT http://www.apple.com/legal/sales_policies/.
Disagree
Agree
Touch here.

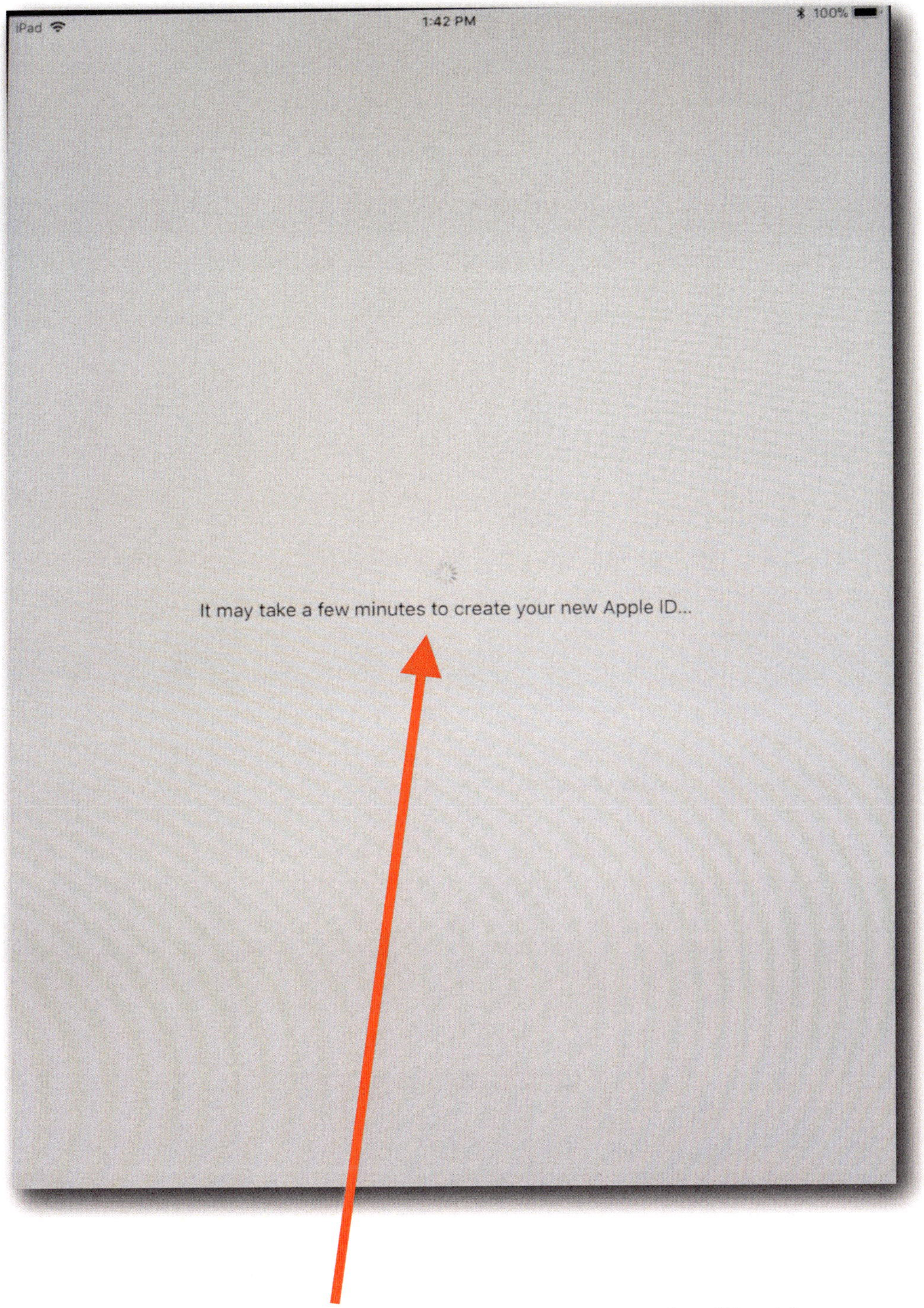
iPad
1:42 PM
100%
It may take a few minutes to create your new Apple ID...

The little spinning wheel appears while the iPad is completing this part of the set up.

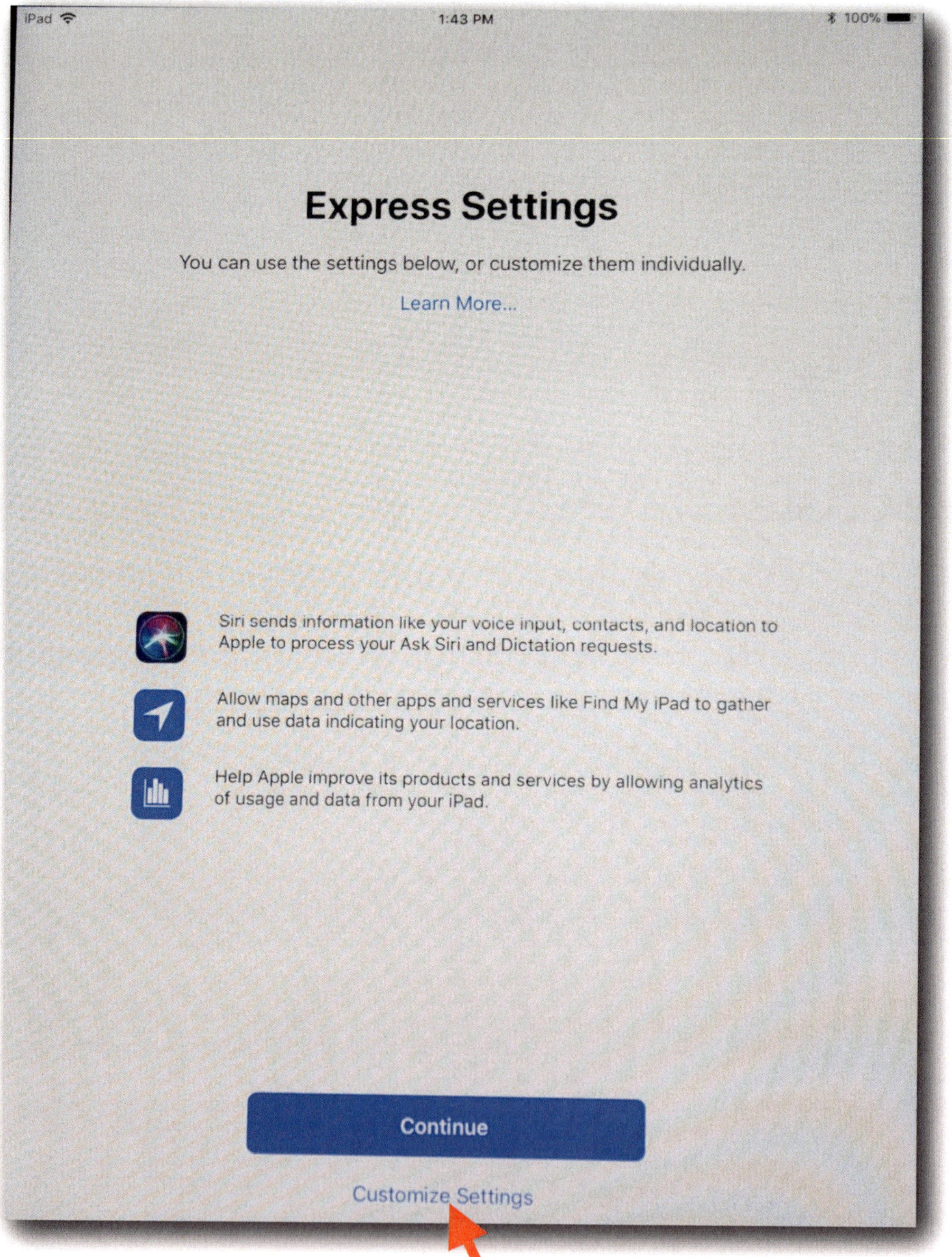
iPad
1:43 PM
100%
Express Settings
You can use the settings below, or customize them individually.
Learn More...
Siri sends information like your voice input, contacts, and location to Apple to process your Ask Siri and Dictation requests.
Allow maps and other apps and services like Find My iPad to gather and use data indicating your location.
Help Apple improve its products and services by allowing analytics of usage and data from your iPad.
Continue
Customize Settings

Touch here.

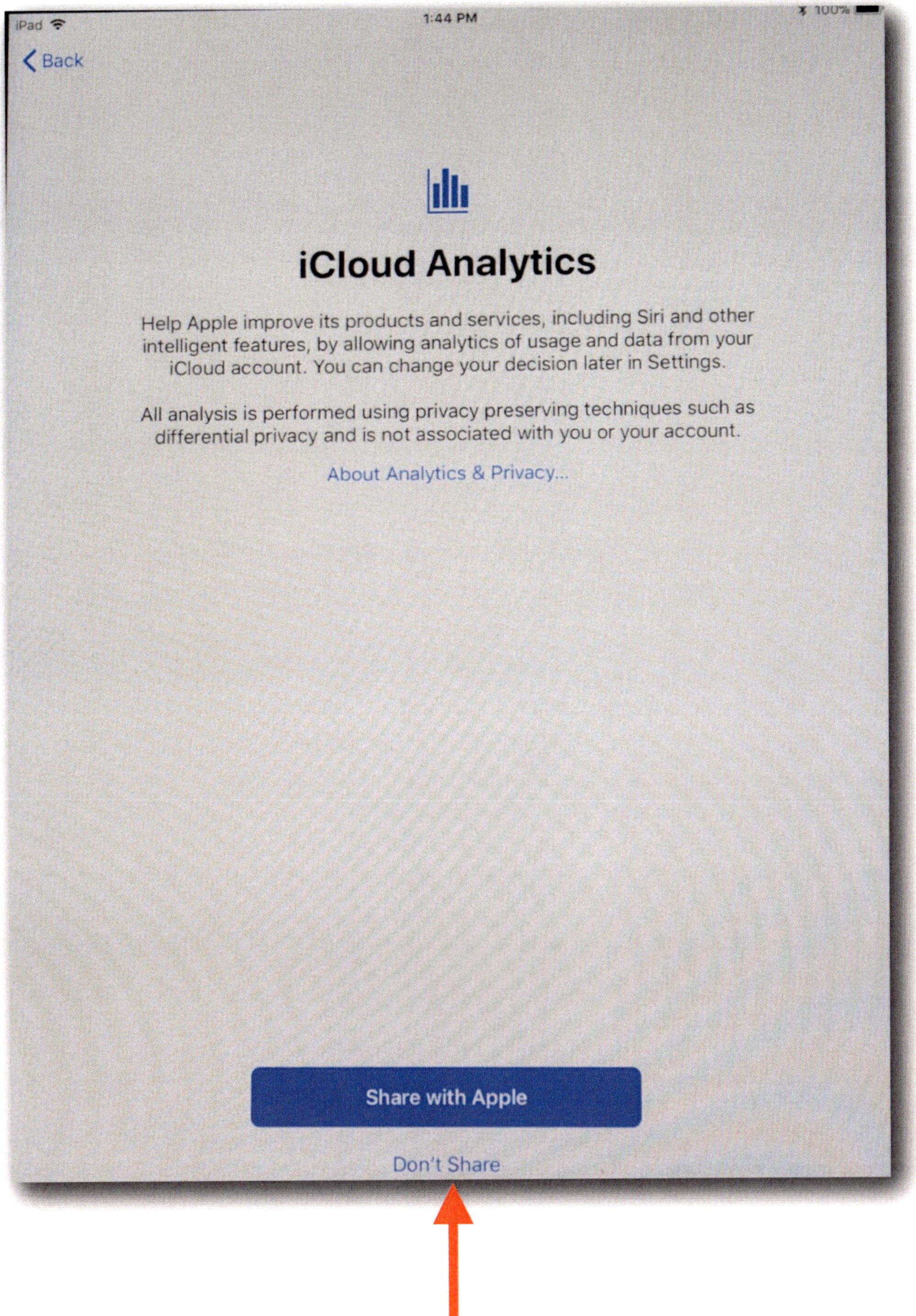
iPad
1:44 PM
100%
Back
iCloud Analytics
Help Apple improve its products and services, including Siri and other intelligent features, by allowing analytics of usage and data from your iCloud account. You can change your decision later in Settings.
All analysis is performed using privacy preserving techniques such as differential privacy and is not associated with you or your account.
About Analytics & Privacy...
Share with Apple
Don't Share

Touch here for now.

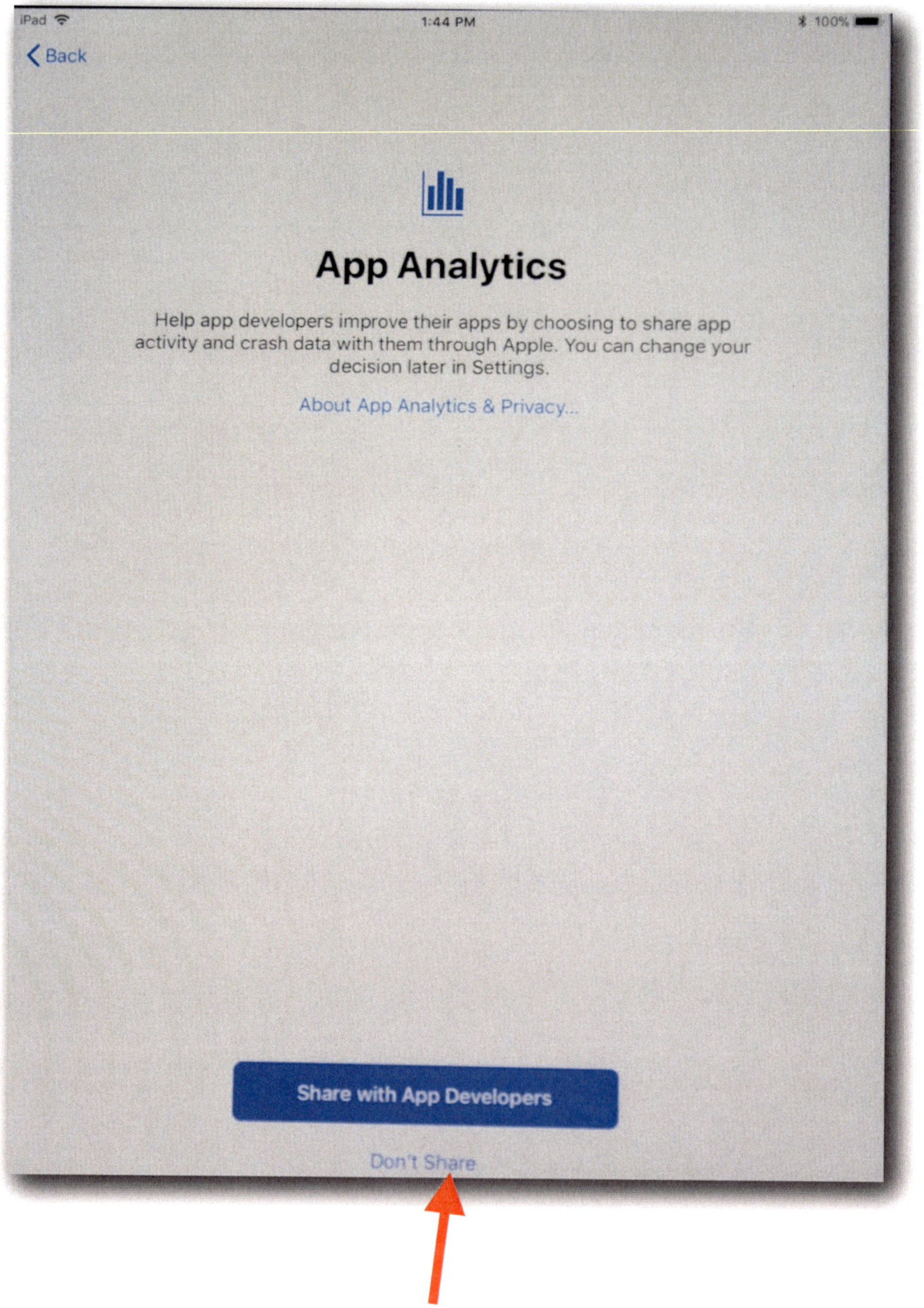
iPad
1:44 PM
100%
Back
App Analytics
Help app developers improve their apps by choosing to share app activity and crash data with them through Apple. You can change your decision later in Settings.
About App Analytics & Privacy...
Share with App Developers
Don't Share

Touch here for now.

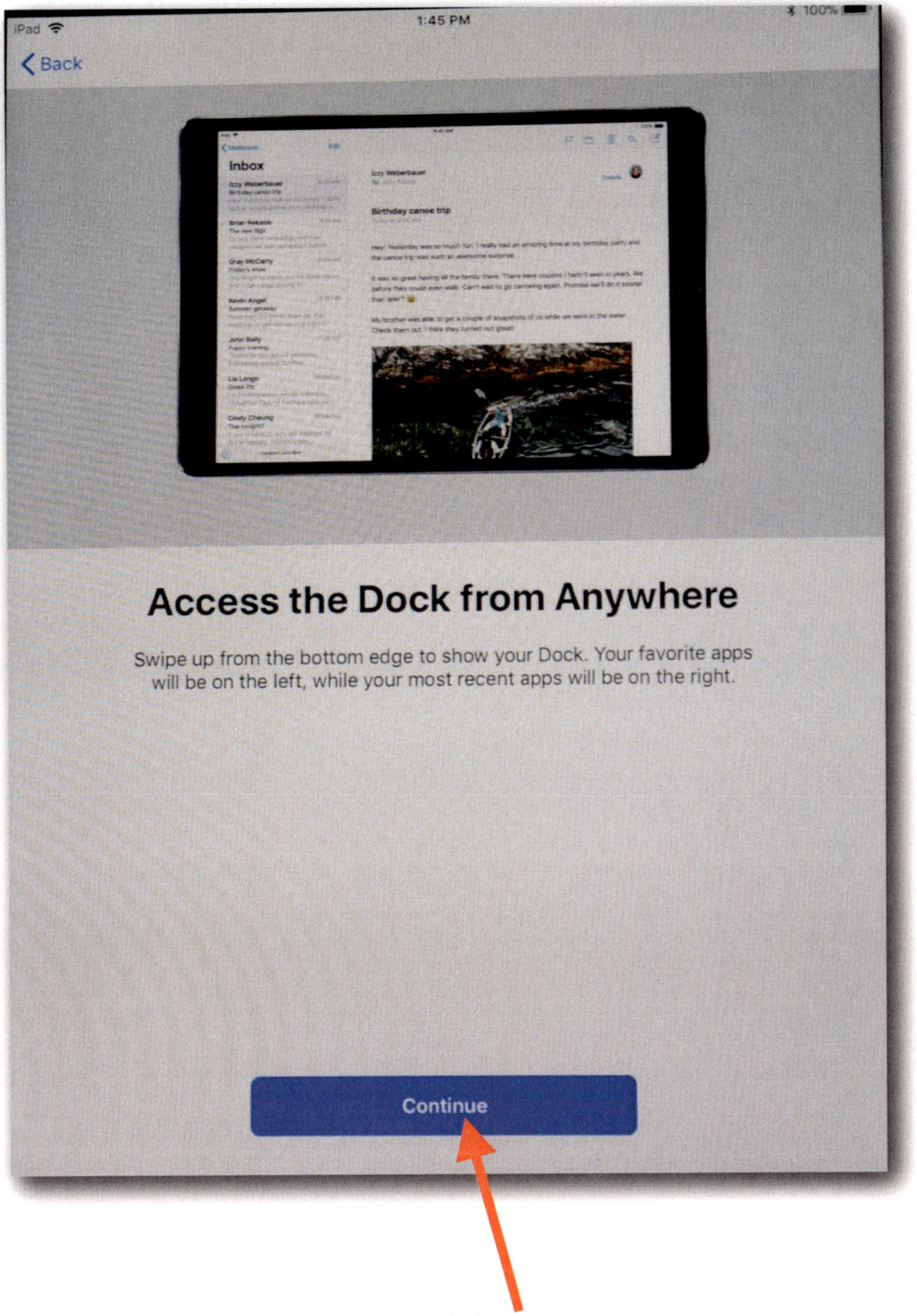
iPad
1:45 PM
100%
Back
Access the Dock from Anywhere
Swipe up from the bottom edge to show your Dock. Your favorite apps will be on the left, while your most recent apps will be on the right.
Continue

Touch here.

iPad
1:45 PM
100%
Back
Switch Between Recent Apps
Swipe up from the bottom edge to reveal the Dock and continue swiping to show the App Switcher.
Continue

Touch here.

Touch here.

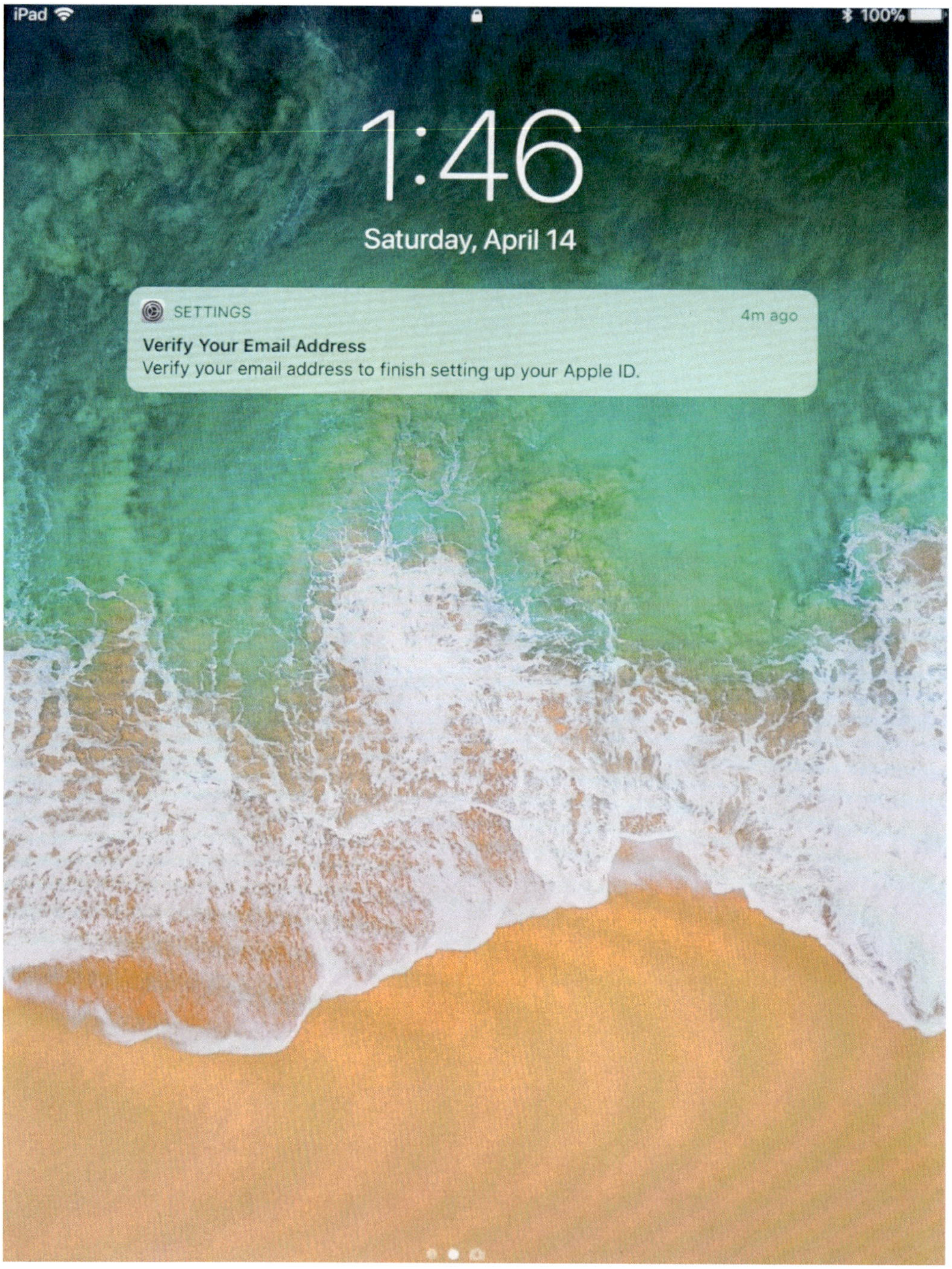
iPad
100%
1:46
Saturday, April 14
SETTINGS
4m ago
Verify Your Email Address
Verify your email address to finish setting up your Apple ID.

Press the Home Button.

You may hold your iPad in Portrait or Landscape view, whichever is more comfortable. You may switch position anytime just by rotating the iPad.

Portrait view

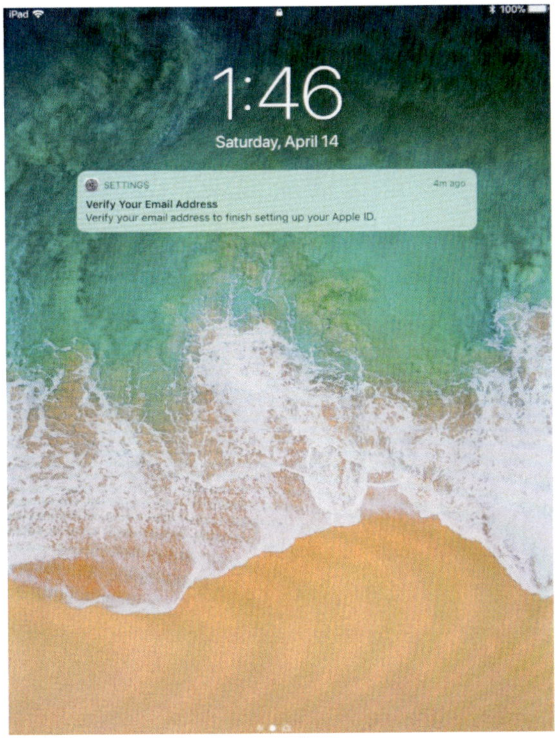

Landscape view

For the remainder of this book images will be in Landscape view.

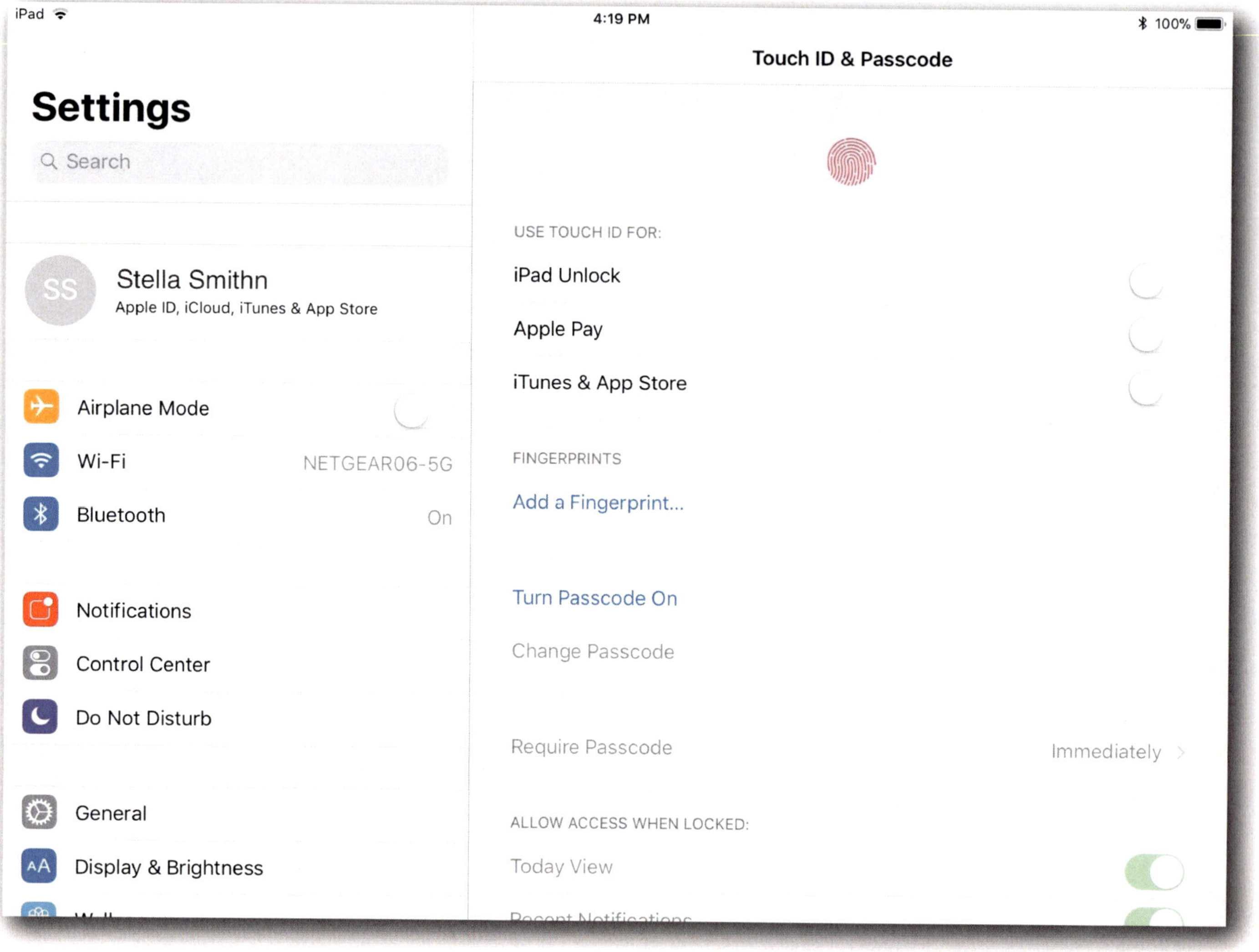

If this page opens, press the Home Button to close.

Welcome to your Home Screen!

Your iPad will look similar to this picture each time you wake it up or turn it on.

These are icons, which represent applications on your iPad. By touching (tapping) an icon, you will open that particular application (app).

This is the Dock.
The most commonly used icons will always be here (such as Text Messaging, Internet, Email.)

The most recently opened apps will appear here.

We're almost done setting up your iPad. Now we're going to increase the text size and make one other adjustment.

Touch here to open settings.

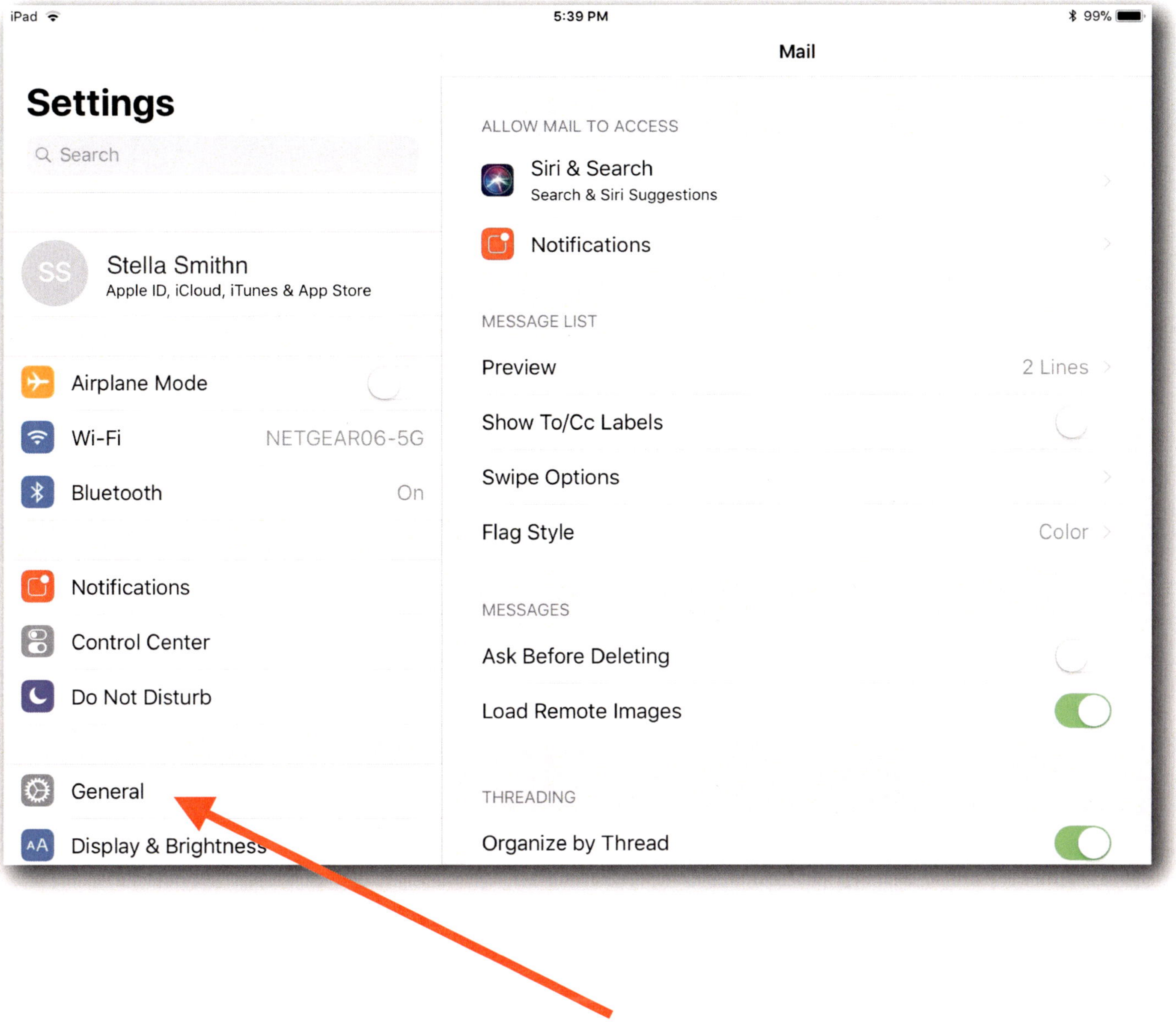
iPad
5:39 PM
99%
Settings
Search
Stella Smithn
Apple ID, iCloud, iTunes & App Store
Airplane Mode
Wi-Fi
NETGEAR06-5G
Bluetooth
On
Notifications
Control Center
Do Not Disturb
General
Display & Brightness
Mail
ALLOW MAIL TO ACCESS
Siri & Search
Search & Siri Suggestions
Notifications
MESSAGE LIST
Preview
2 Lines
Show To/Cc Labels
Swipe Options
Flag Style
Color
MESSAGES
Ask Before Deleting
Load Remote Images
THREADING
Organize by Thread

Touch here.

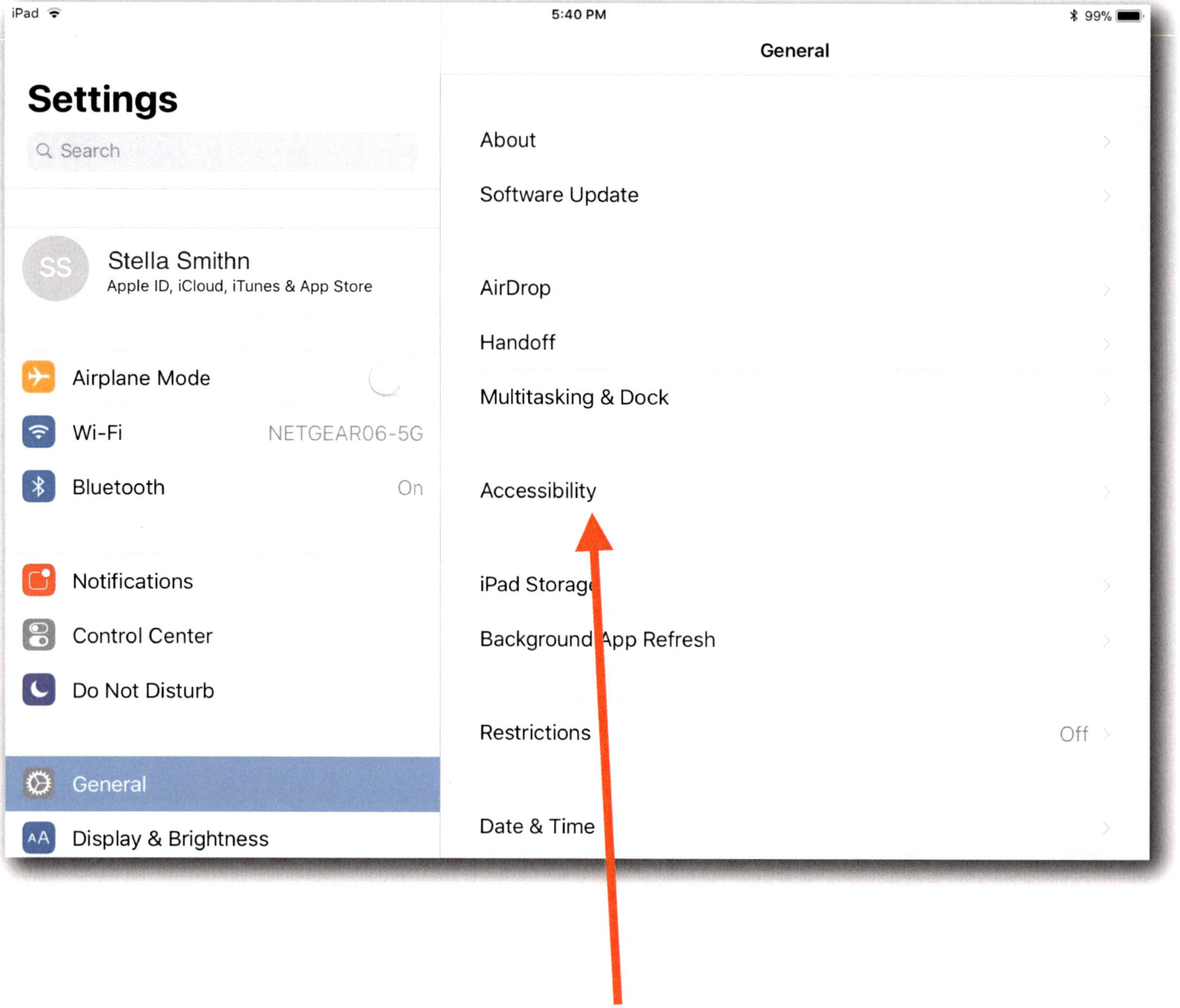

iPad
5:40 PM
99%
Settings
Search
SS
Stella Smithn
Apple ID, iCloud, iTunes & App Store
Airplane Mode
Wi-Fi
NETGEAR06-5G
Bluetooth
On
Notifications
Control Center
Do Not Disturb
General
AA
Display & Brightness
General
About
Software Update
AirDrop
Handoff
Multitasking & Dock
Accessibility
iPad Storage
Background App Refresh
Restrictions
Off
Date & Time

Touch here.

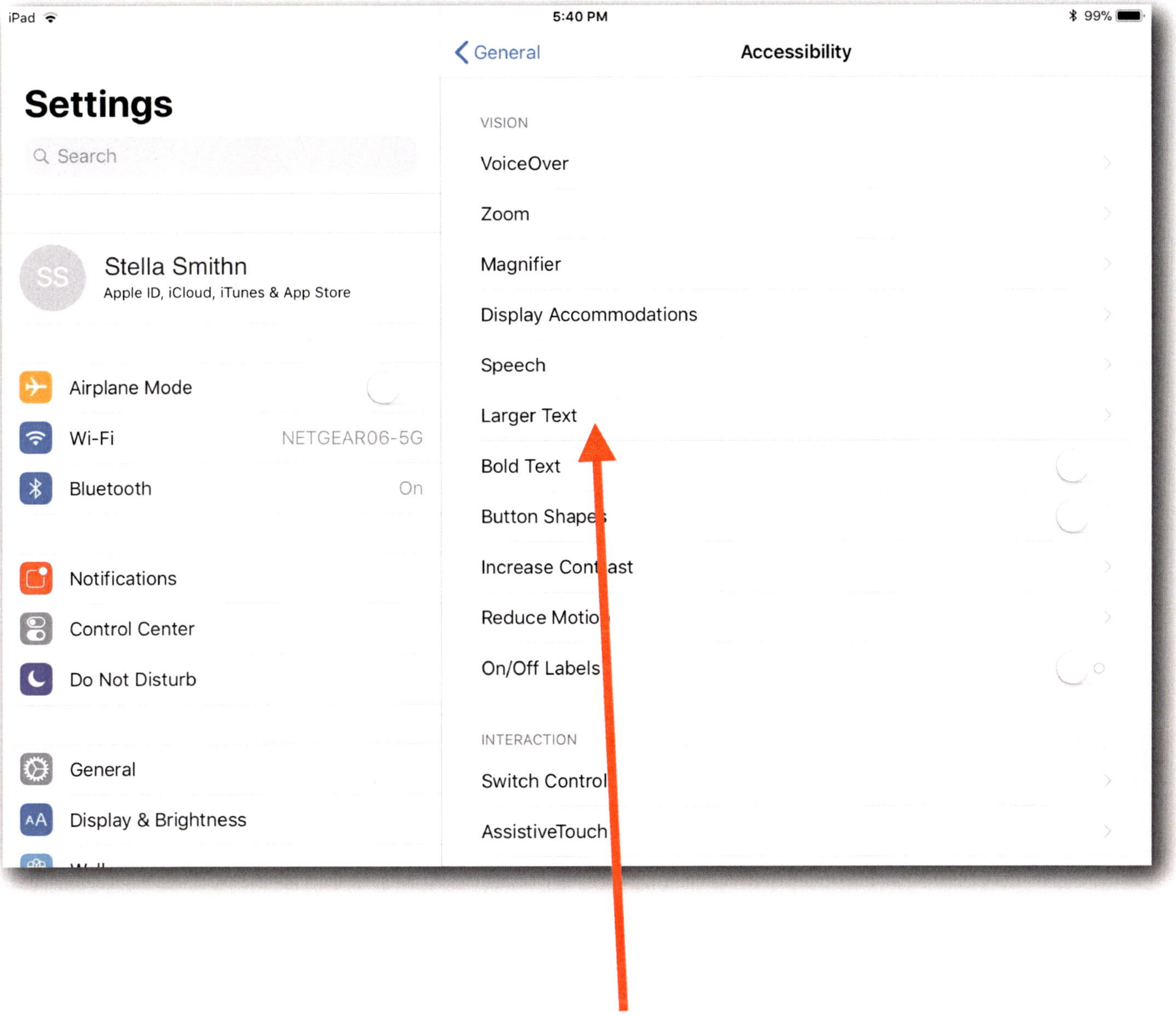
iPad
5:40 PM
99%
Settings
Search
SS
Stella Smithn
Apple ID, iCloud, iTunes & App Store
Airplane Mode
Wi-Fi
NETGEAR06-5G
Bluetooth
On
Notifications
Control Center
Do Not Disturb
General
Display & Brightness
General
Accessibility
VISION
VoiceOver
Zoom
Magnifier
Display Accommodations
Speech
Larger Text
Bold Text
Button Shapes
Increase Contrast
Reduce Motion
On/Off Labels
INTERACTION
Switch Control
AssistiveTouch

Touch here.

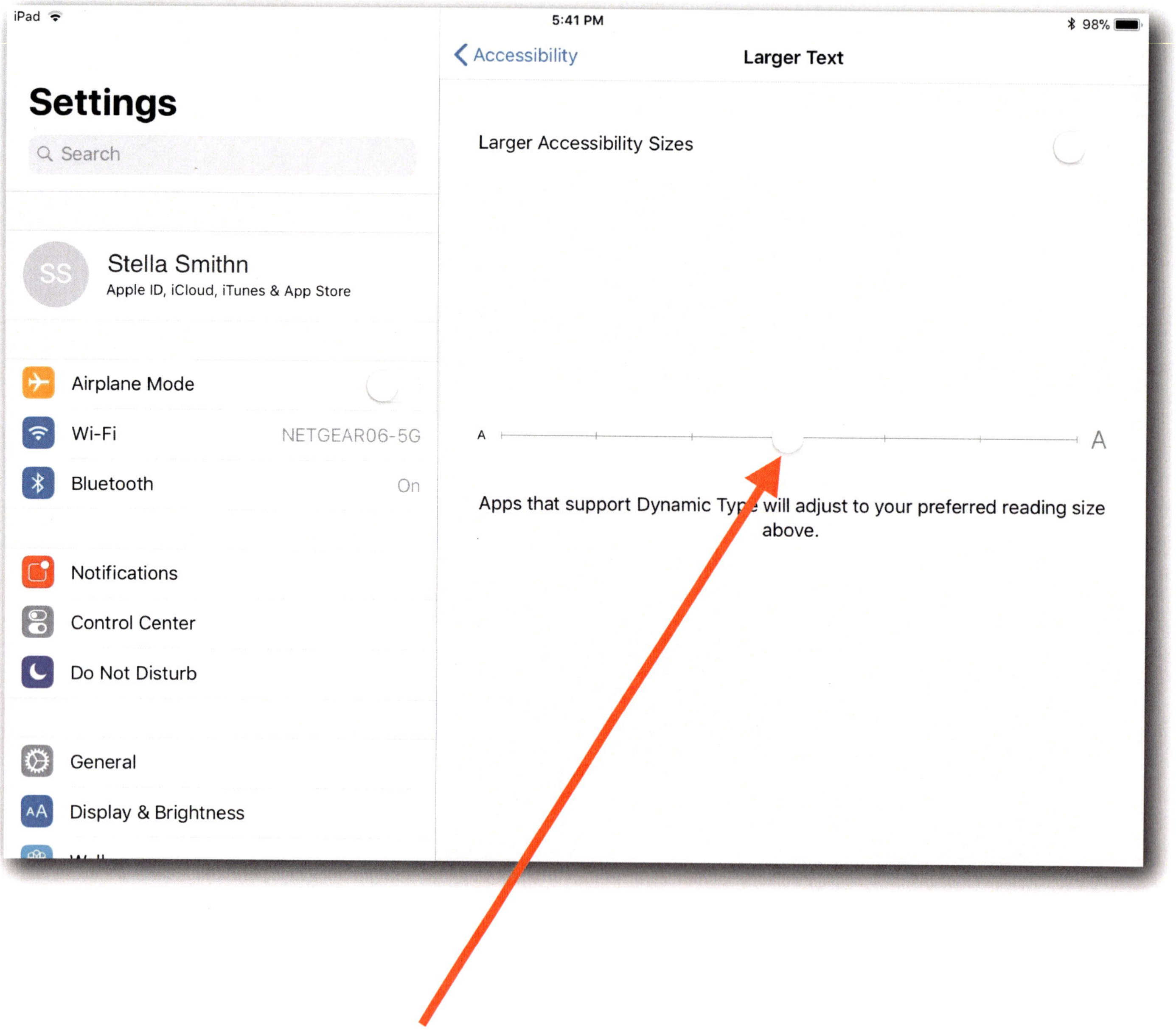

Touch and hold this button. Slide it to the right to increase text size. You may increase or decrease text to any size you choose.

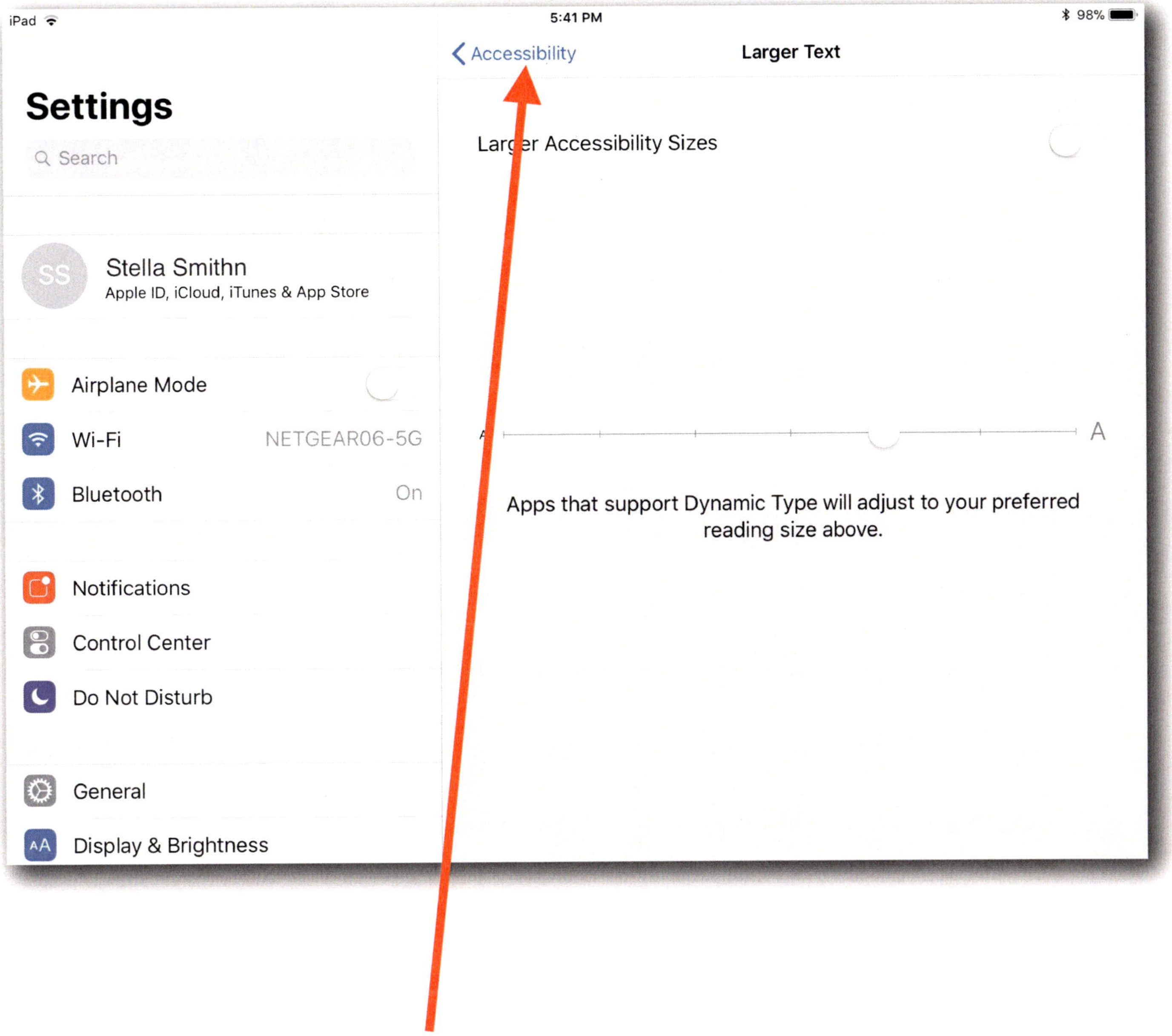

Touch here to go back to previous page.

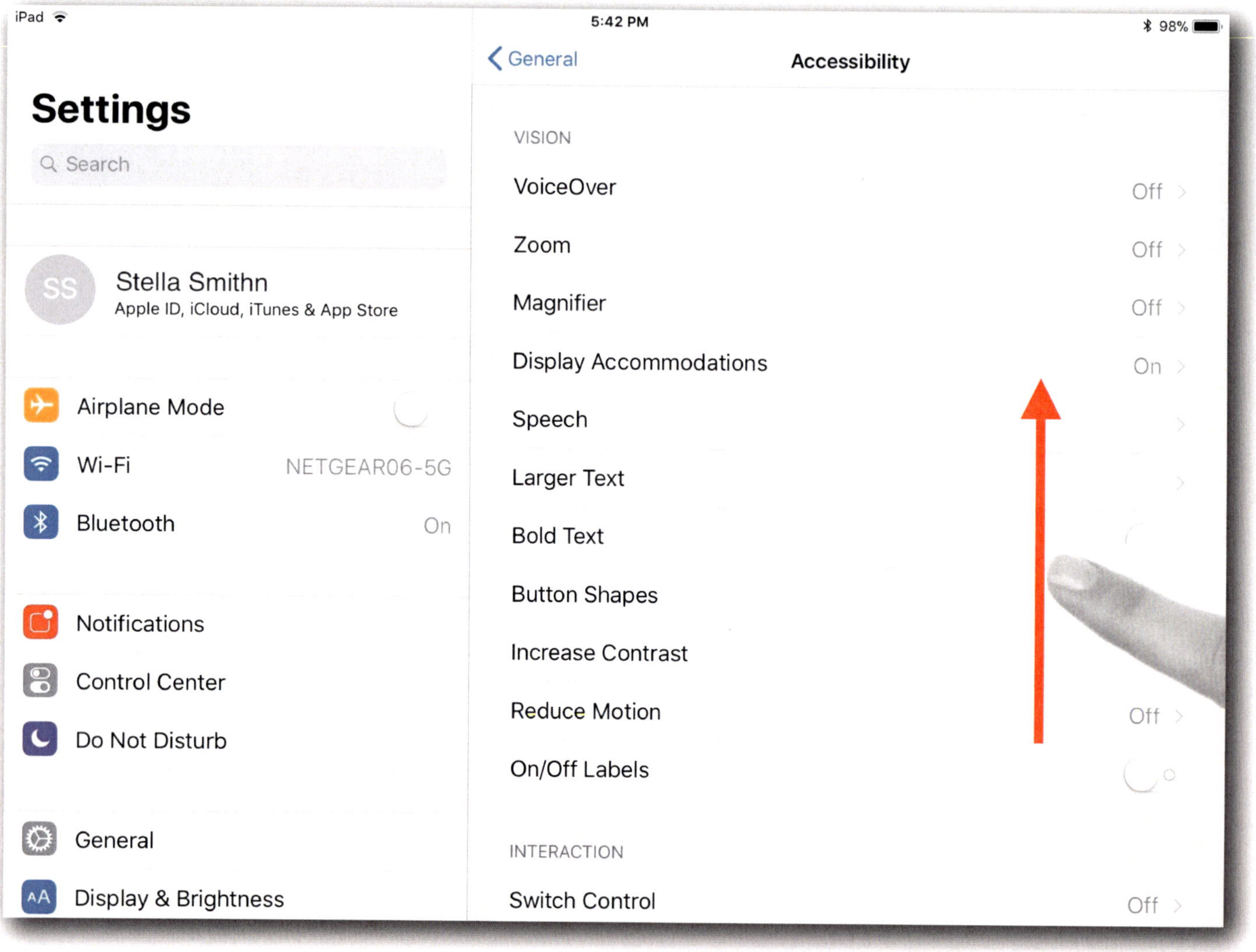

Scroll down the page by gently sliding your finger up the screen.

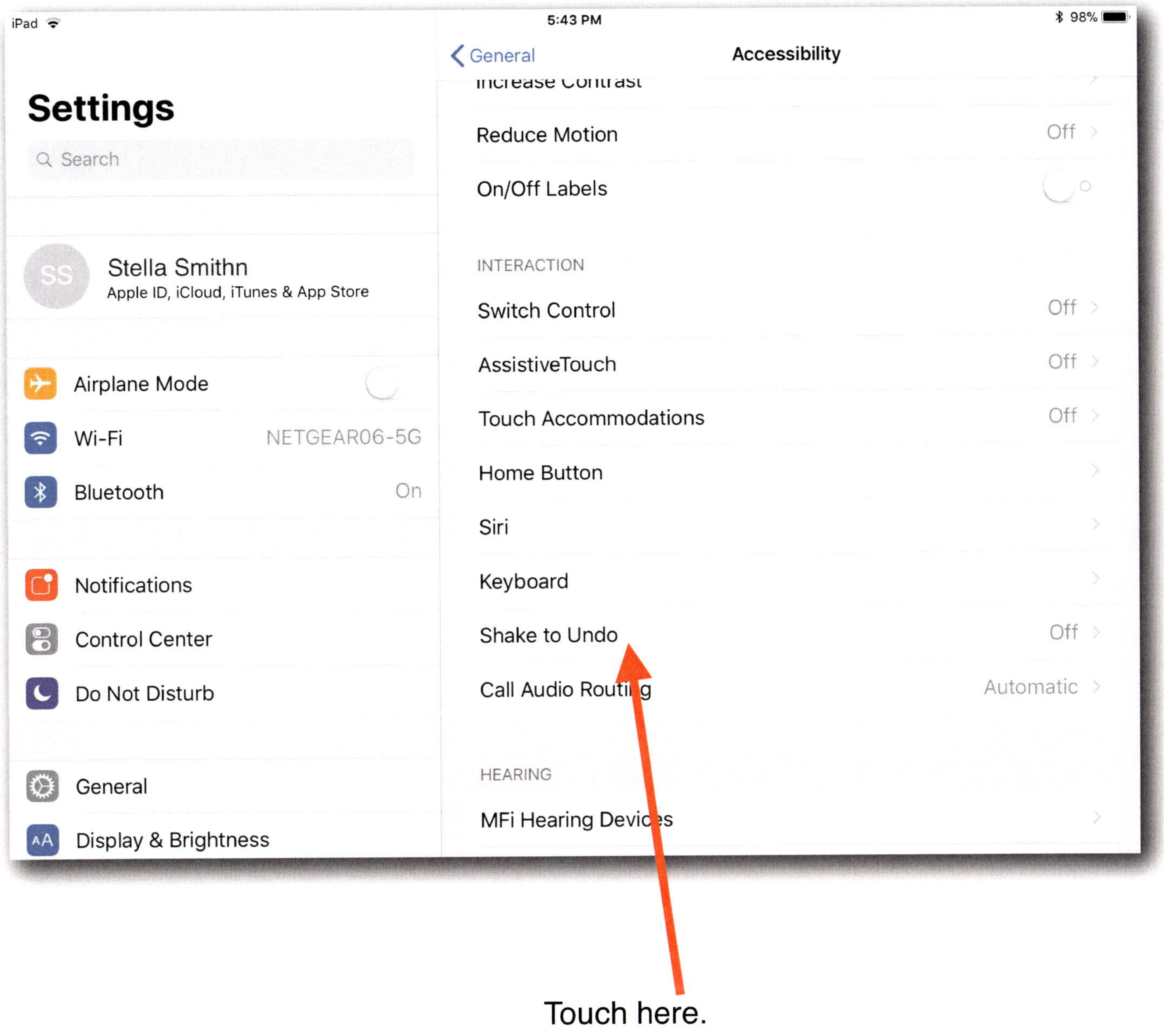

iPad
5:43 PM
98%
Settings
Search
SS
Stella Smithn
Apple ID, iCloud, iTunes & App Store
Airplane Mode
Wi-Fi
NETGEAR06-5G
Bluetooth
On
Notifications
Control Center
Do Not Disturb
General
Display & Brightness
General
Accessibility
Reduce Motion
Off
On/Off Labels
INTERACTION
Switch Control
Off
AssistiveTouch
Off
Touch Accommodations
Off
Home Button
Siri
Keyboard
Shake to Undo
Off
Call Audio Routing
Automatic
HEARING
MFi Hearing Devices
Touch here.

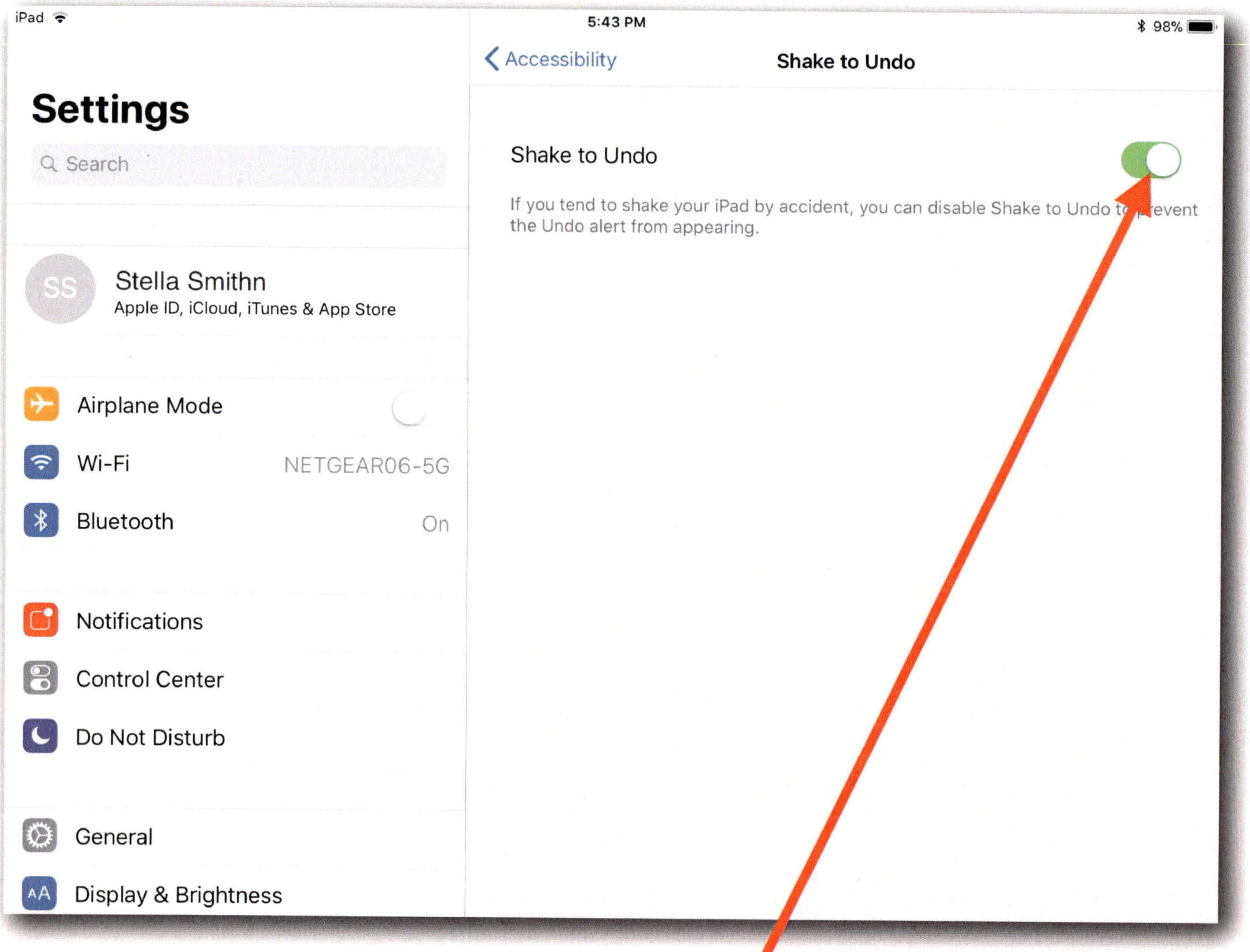
iPad
5:43 PM
98%
Settings
Search
Stella Smithn
Apple ID, iCloud, iTunes & App Store
Airplane Mode
Wi-Fi
NETGEAR06-5G
Bluetooth
On
Notifications
Control Center
Do Not Disturb
General
Display & Brightness
Accessibility
Shake to Undo
Shake to Undo
If you tend to shake your iPad by accident, you can disable Shake to Undo to prevent the Undo alert from appearing.

Touch here.

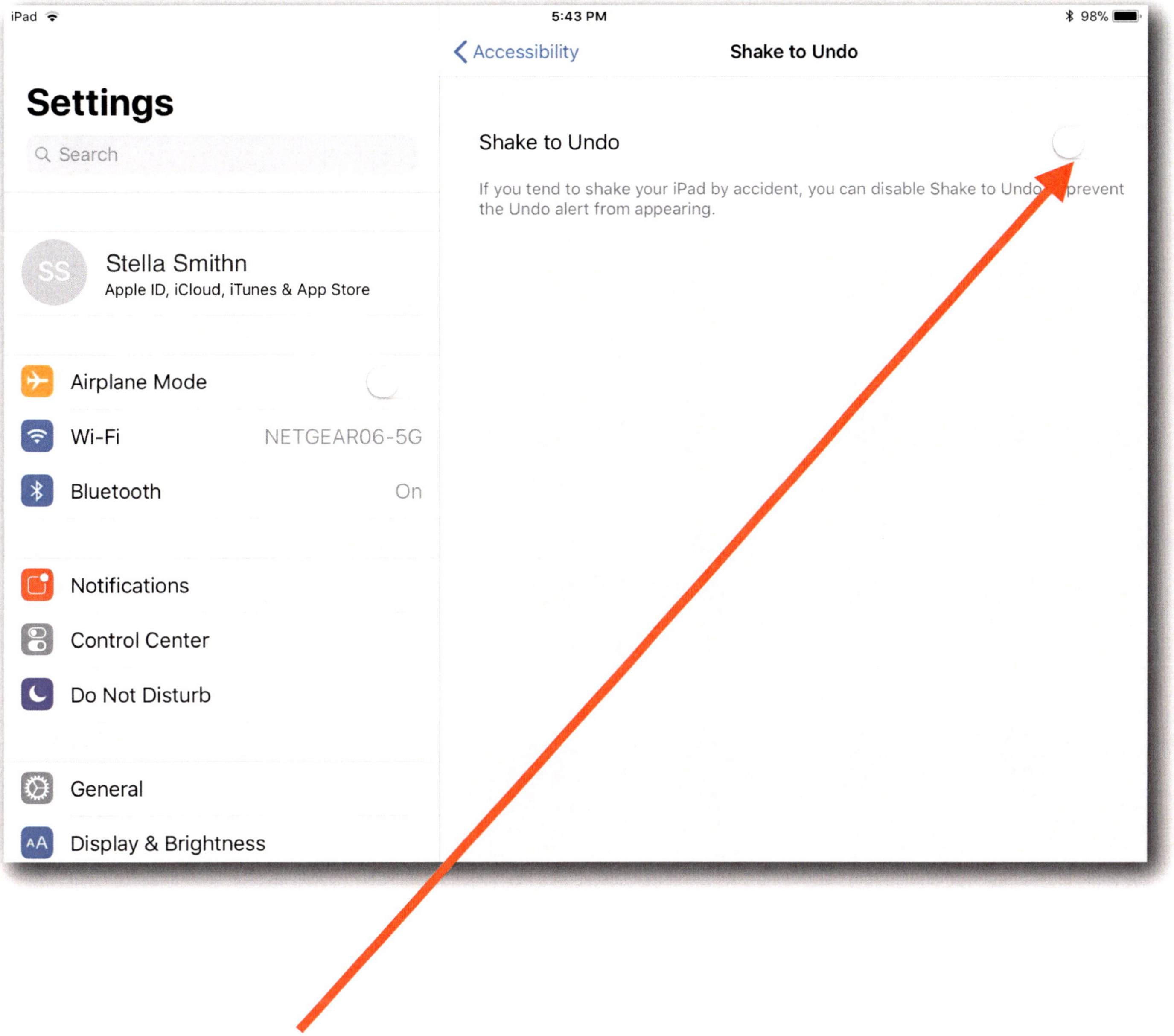

When the button is no longer green, this feature has been turned off.

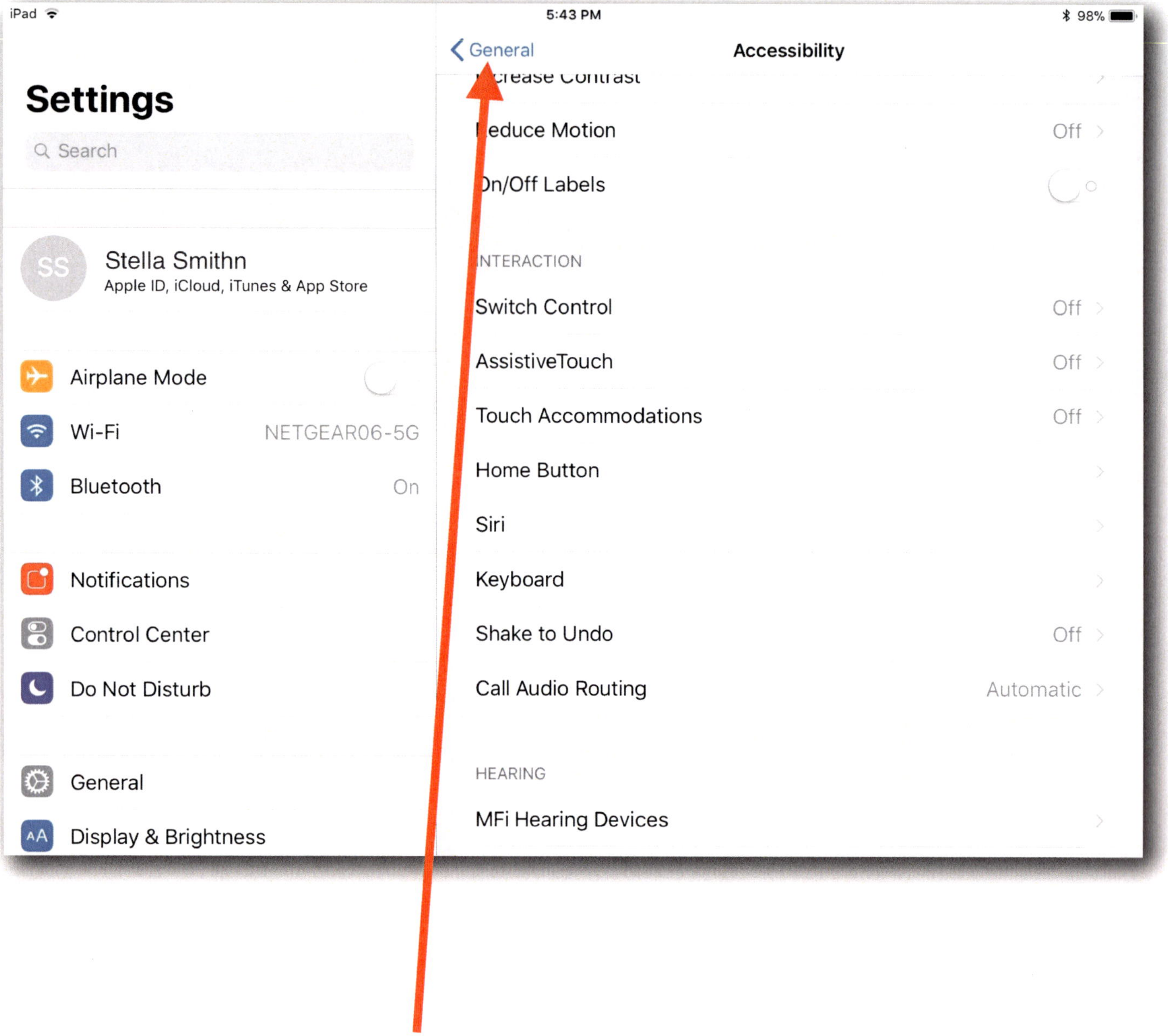

Touch here to go back to previous page.

Press the Home Button to close Settings.

Setting Up Email

Touch here to open email.

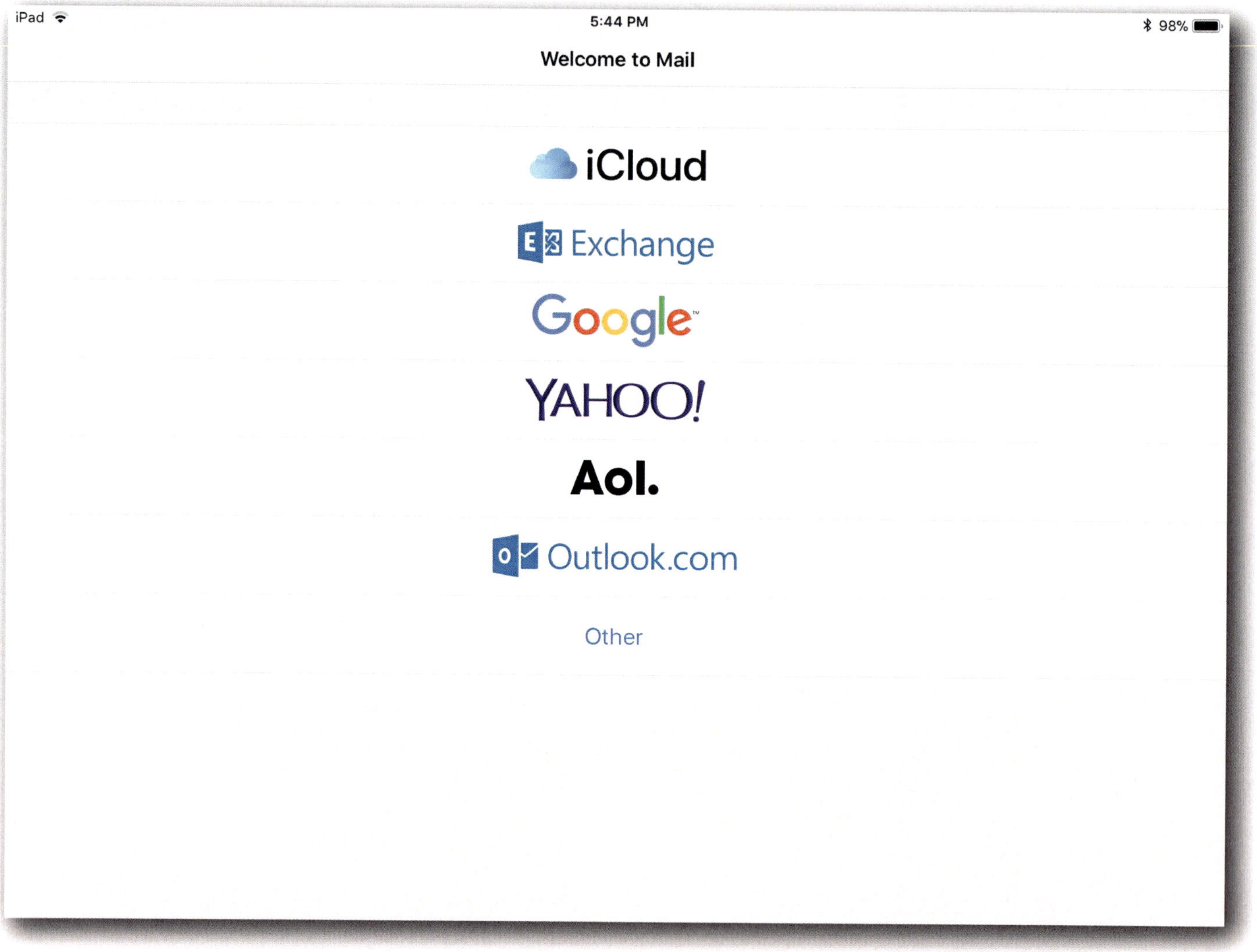

Touch the name of your email server.

For this step I will be using Gmail, (which is Google) to demonstrate.

Note: If you don't already have an email address, you can use your iCloud email, which you may have set up earlier in the set-up process, or you can create a new email for free on one of the email servers listed above. Just touch your choice from the list and follow the prompts.

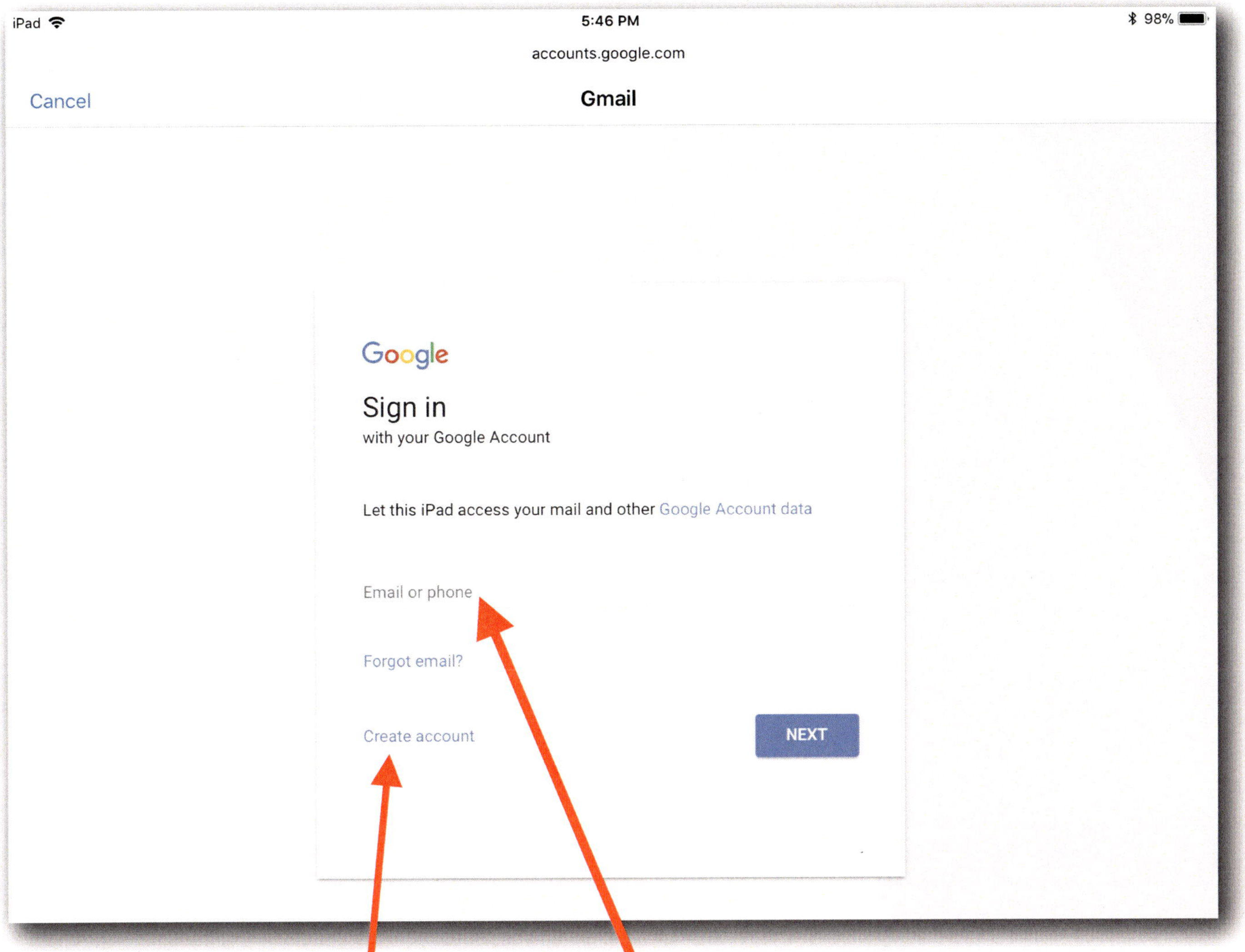

Touch here to sign in with your existing Gmail account.

OR...

Touch here to create a new account and follow the prompts.

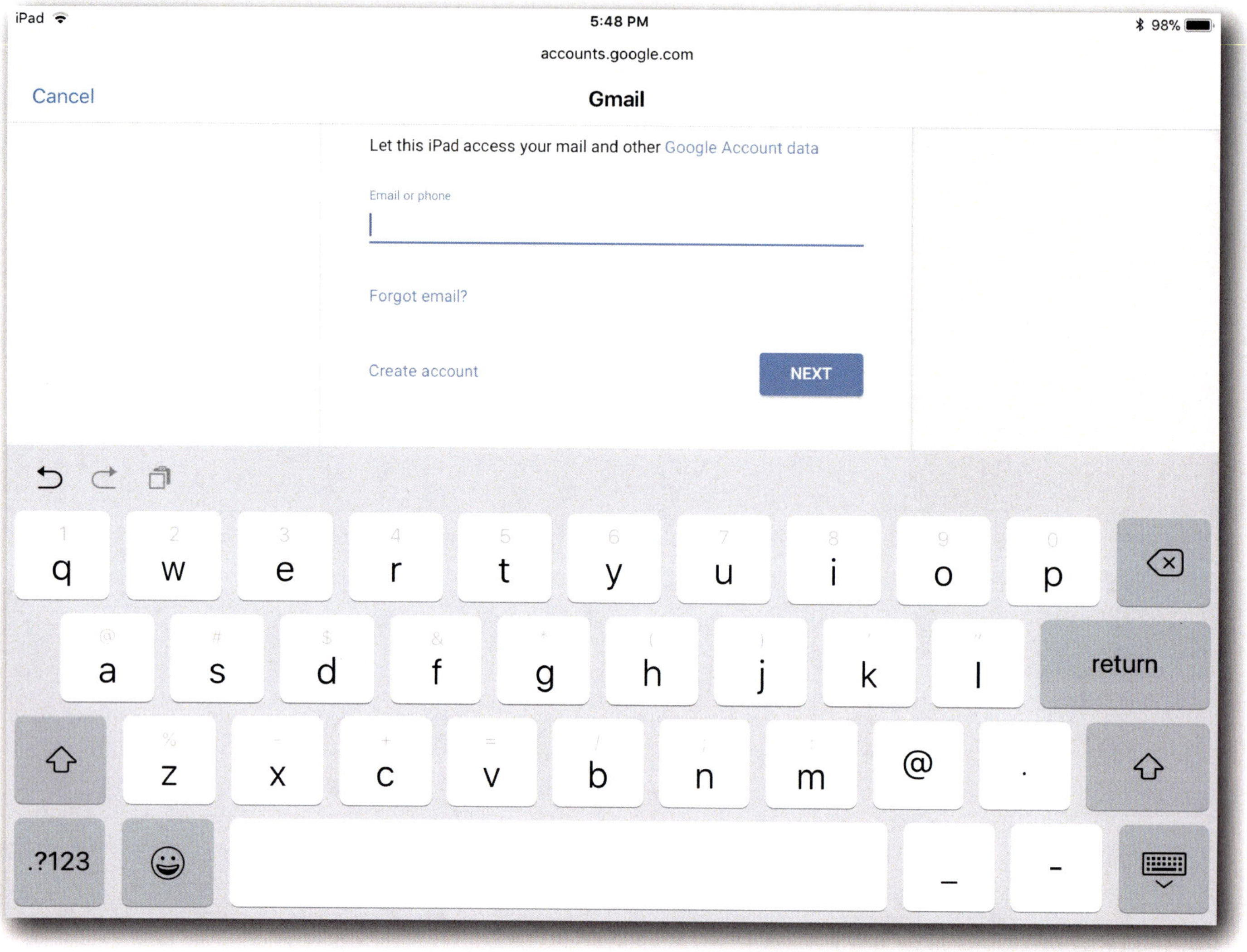

Type your email address.

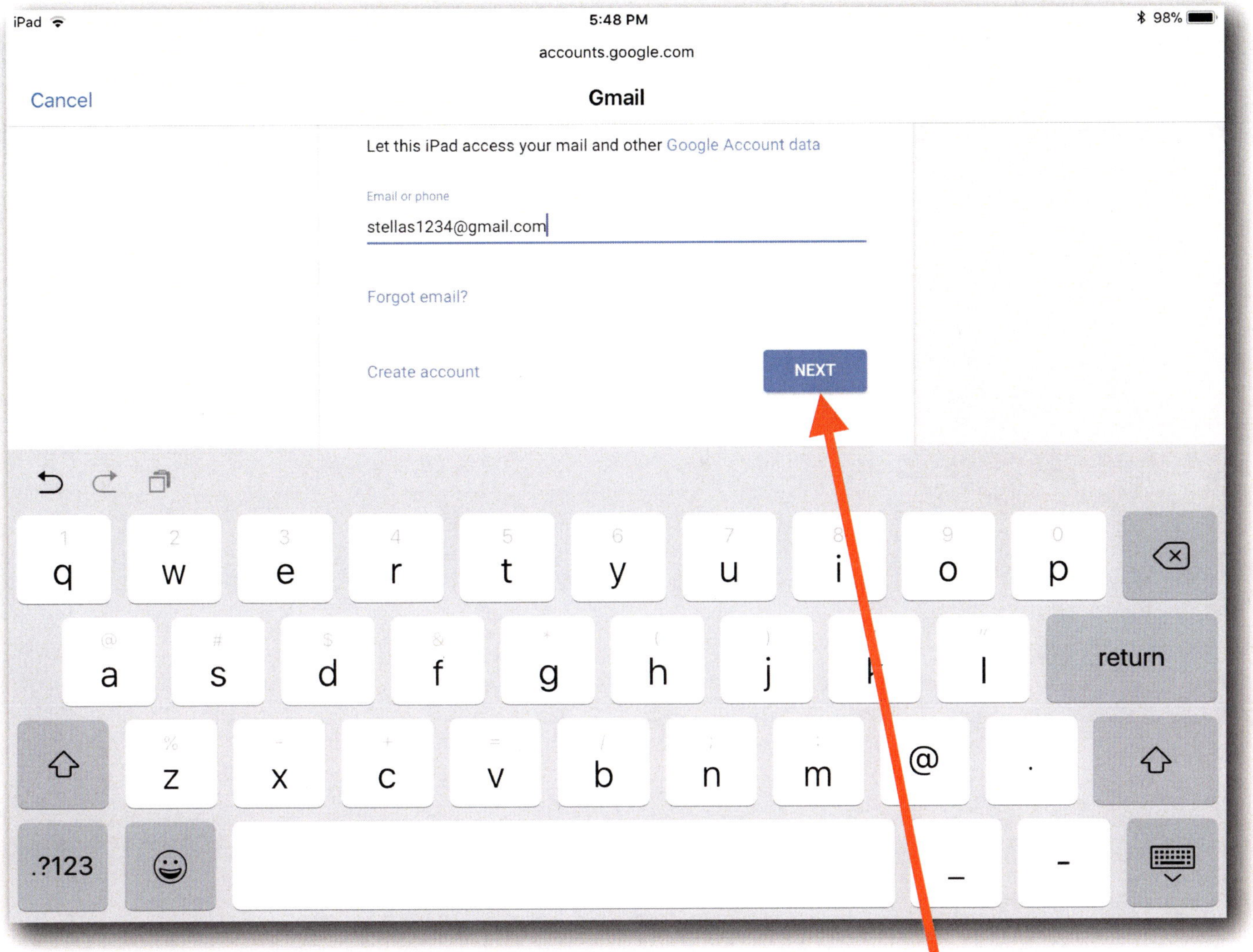

When finished typing your email address, touch here.

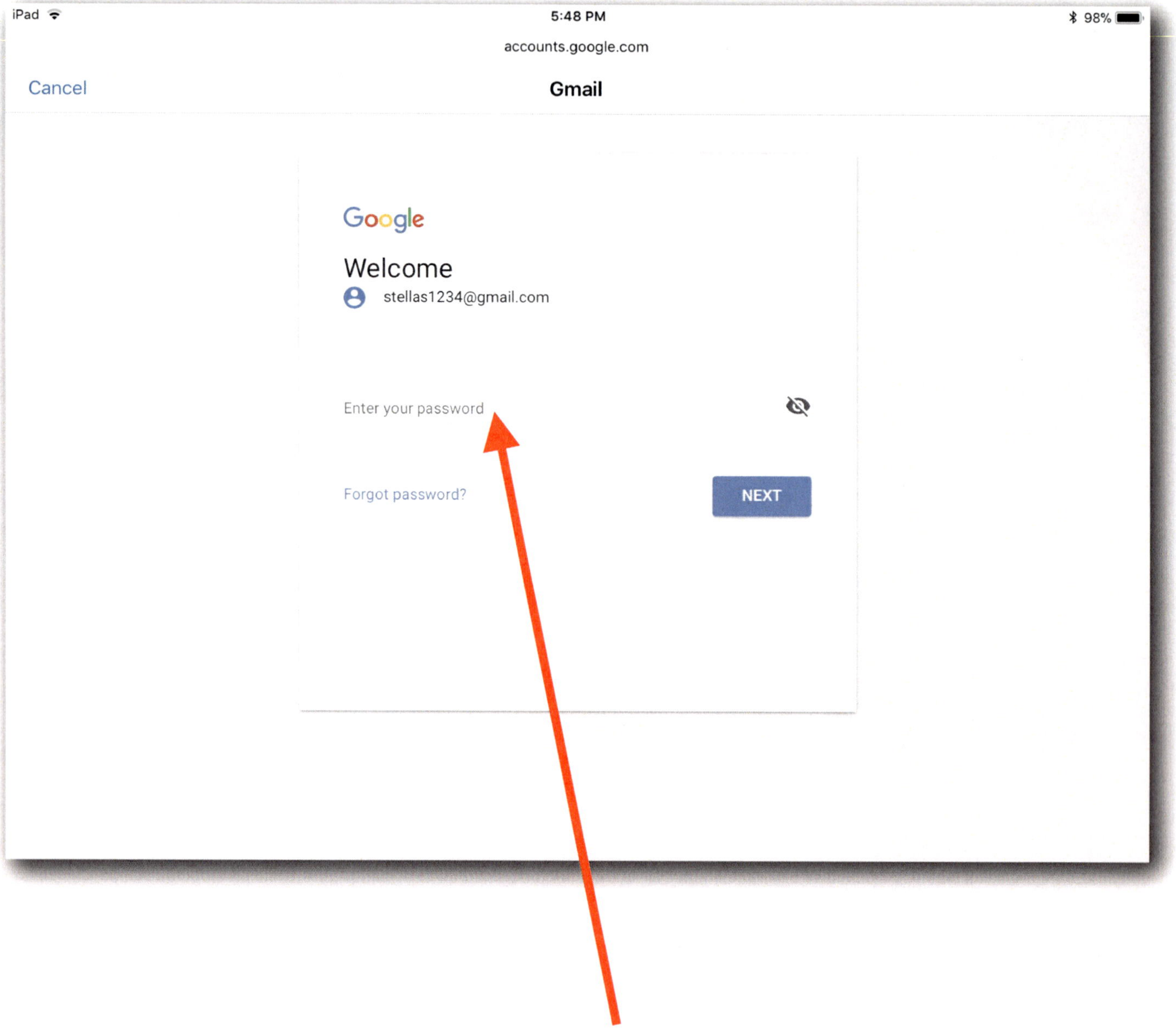
iPad
5:48 PM
98%
accounts.google.com
Cancel
Gmail
Google
Welcome
stellas1234@gmail.com
Enter your password
Forgot password?
NEXT

Touch here.

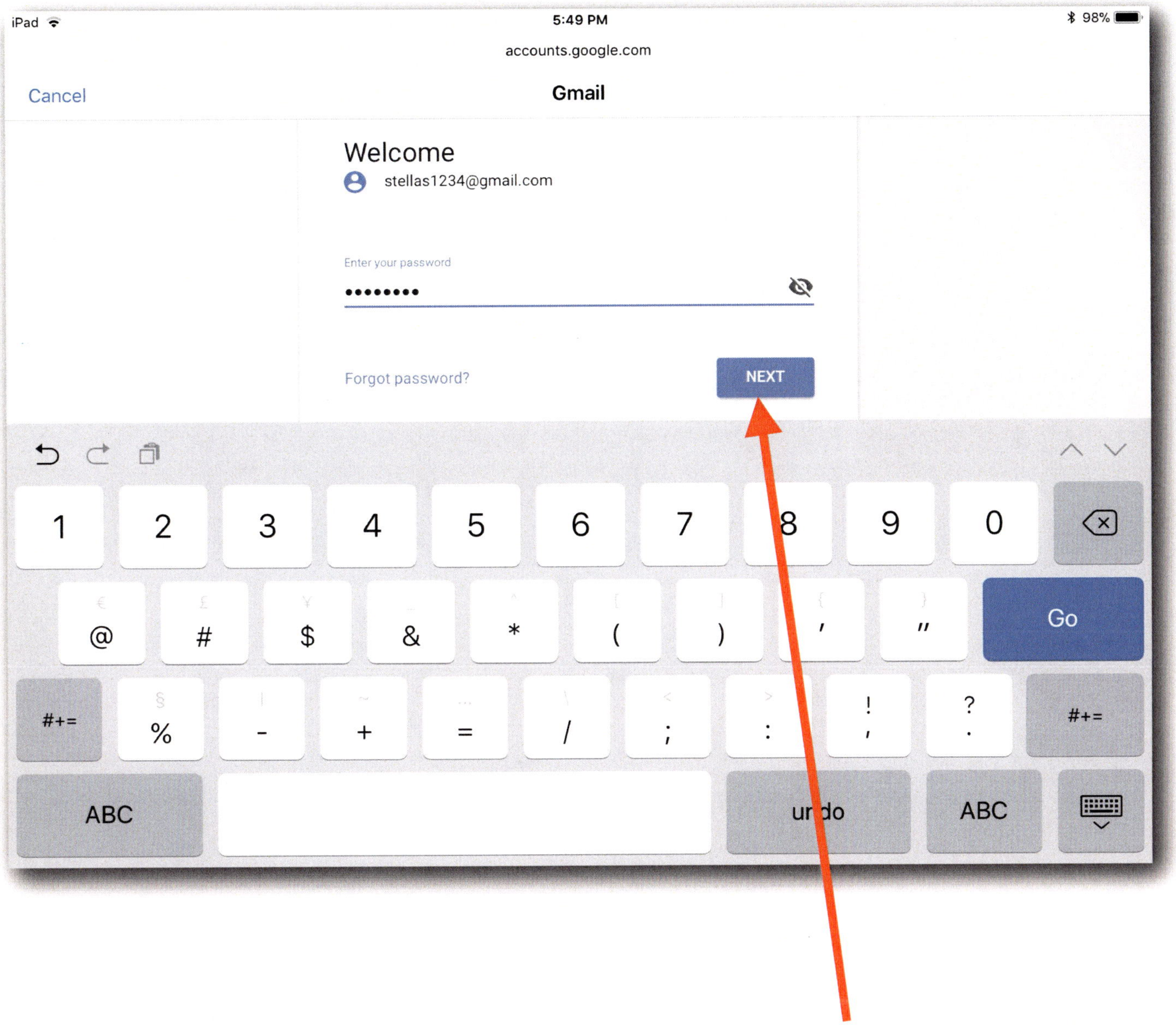

Type your email password. Then touch here.

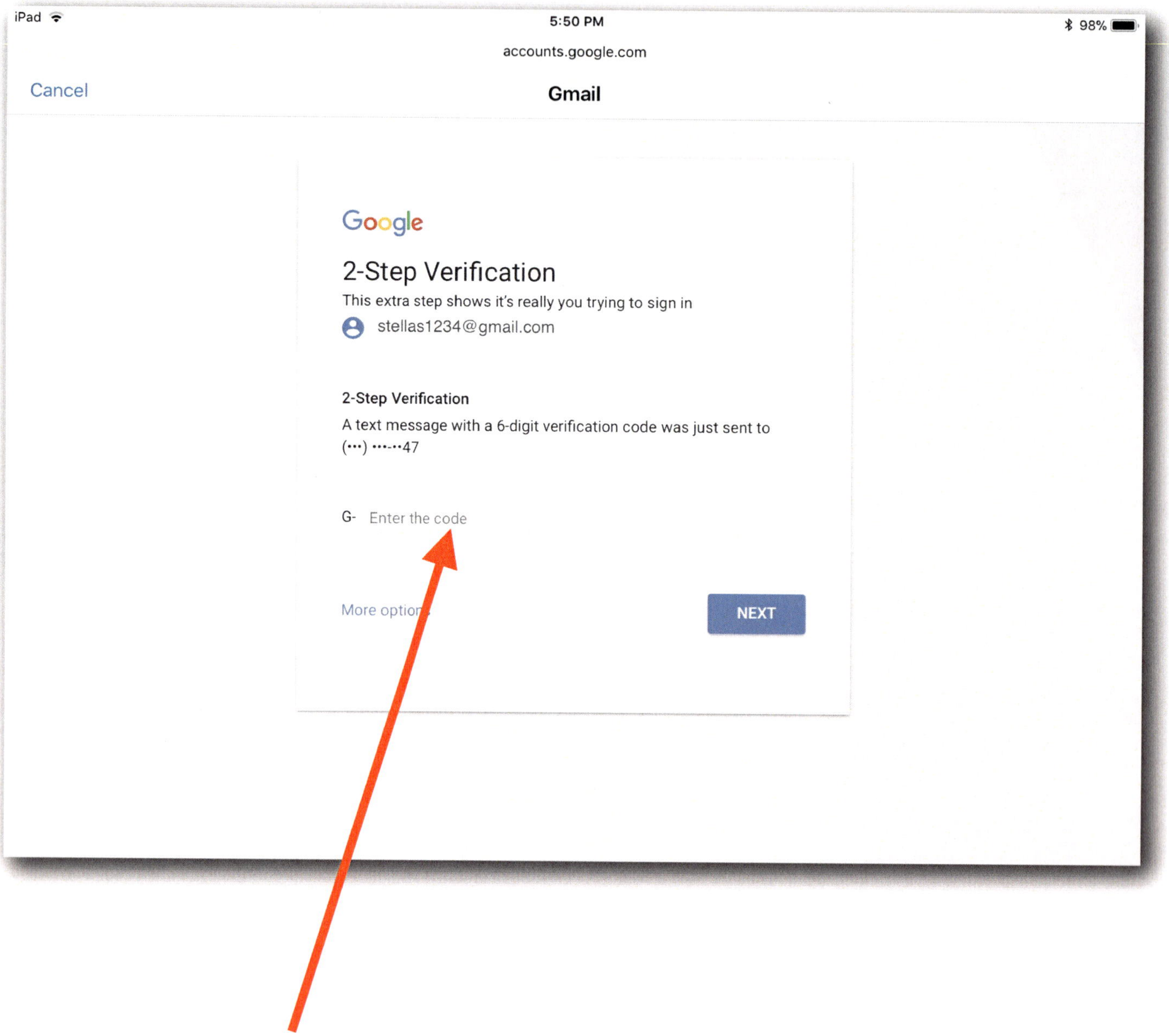
iPad
5:50 PM
98%
accounts.google.com
Cancel
Gmail
Google
2-Step Verification
This extra step shows it's really you trying to sign in
stellas1234@gmail.com
2-Step Verification
A text message with a 6-digit verification code was just sent to (•••) •••-••47
G- Enter the code
NEXT

Touch here and type in the code that was sent to your phone.

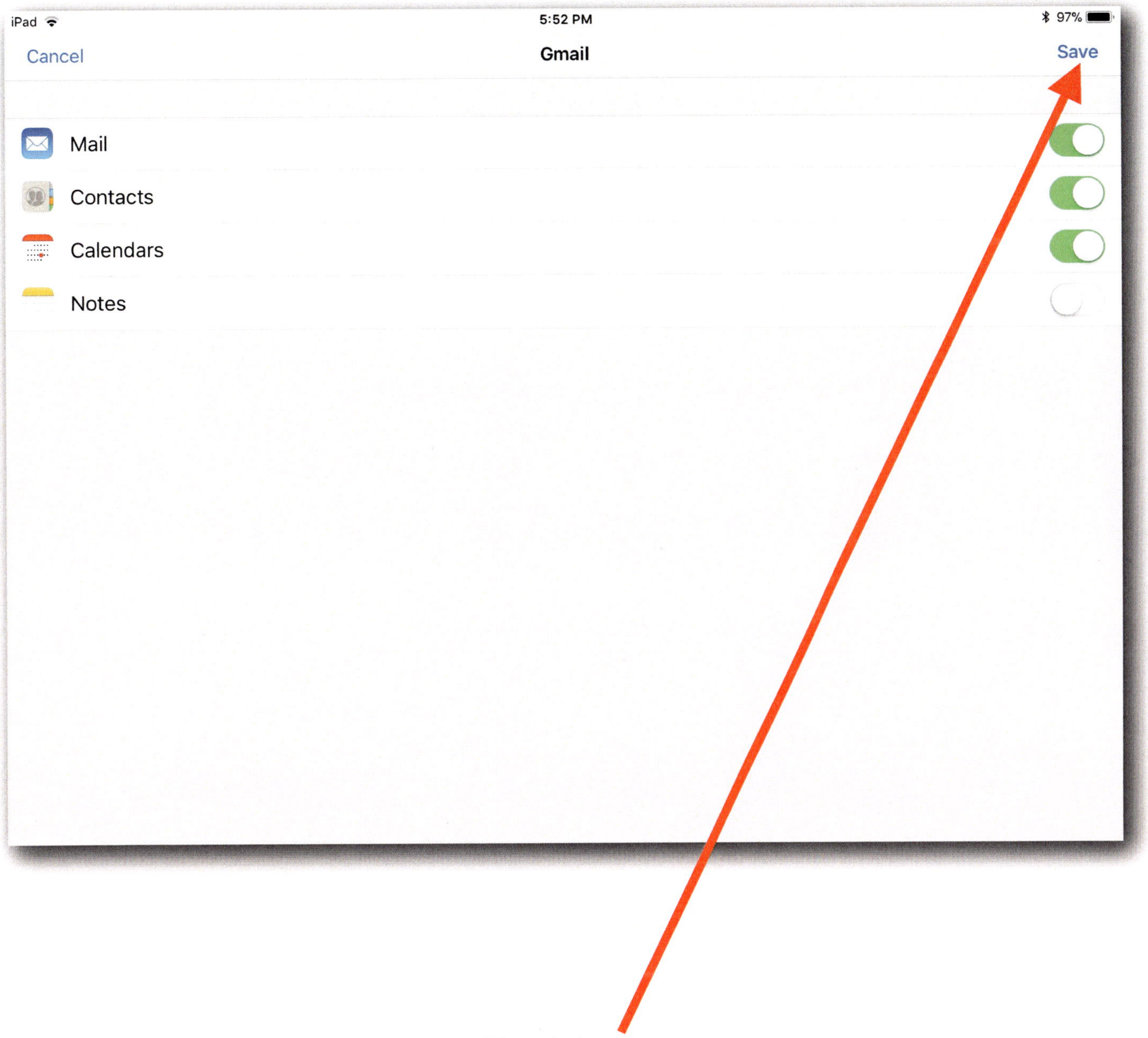
iPad
5:52 PM
97%
Cancel
Gmail
Save
Mail
Contacts
Calendars
Notes

Touch here.

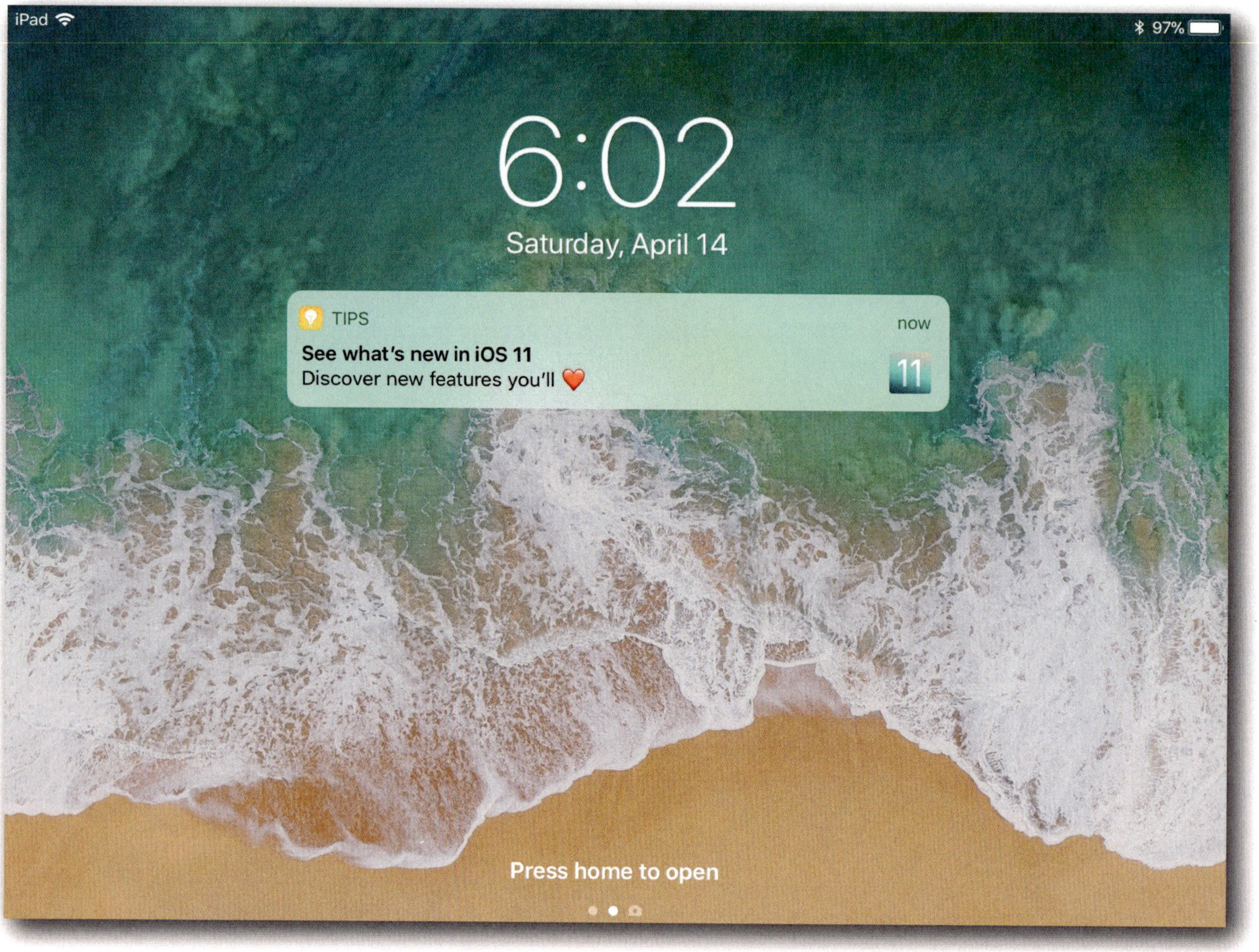

Press the Home Button to continue.

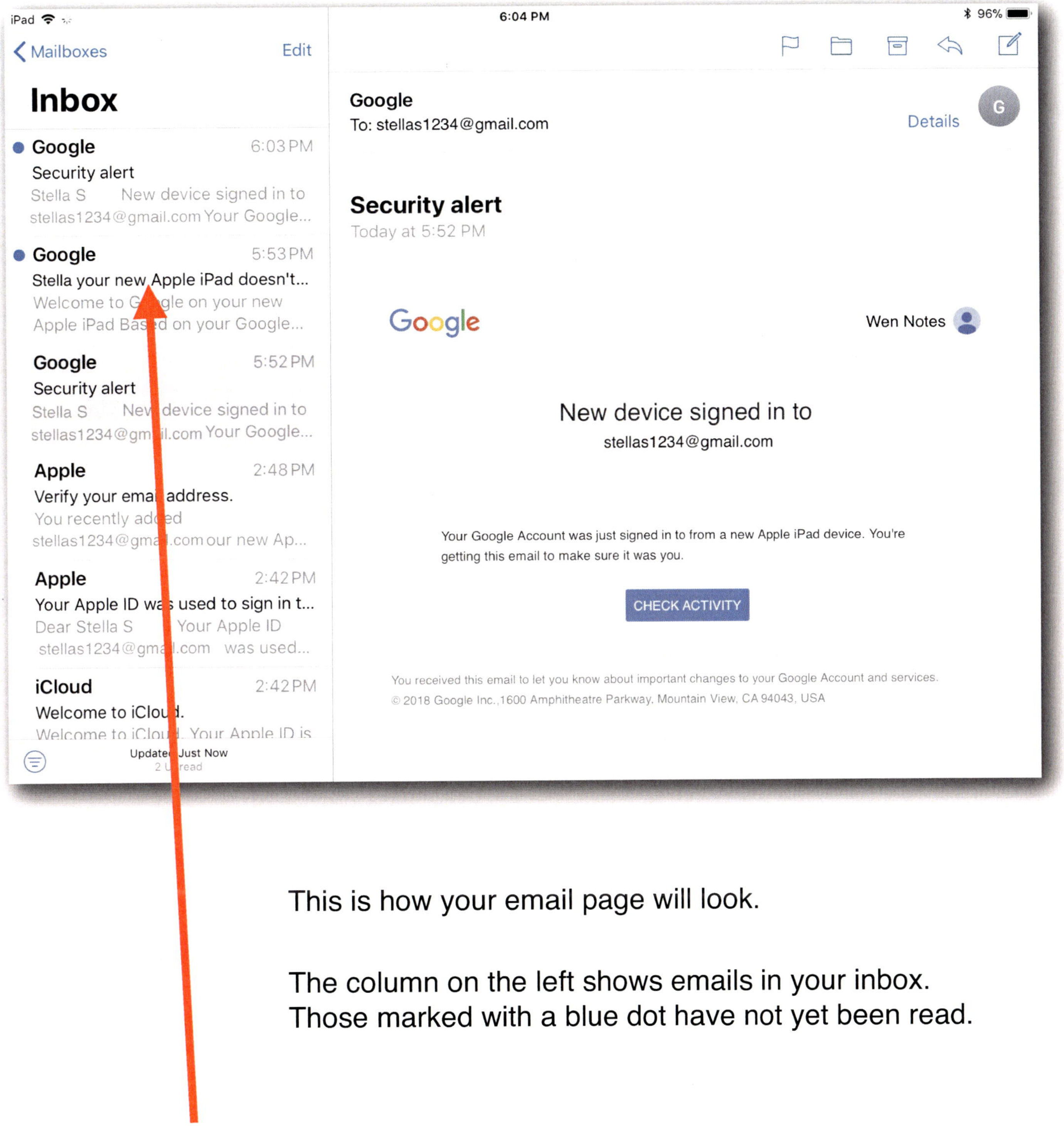
iPad
6:04 PM
96%
Mailboxes
Edit
Inbox
Google
6:03 PM
Security alert
Stella S New device signed in to
stellas1234@gmail.com Your Google...
Google
5:53 PM
Stella your new Apple iPad doesn't...
Google
5:52 PM
Security alert
Apple
2:48 PM
Verify your email address.
Apple
2:42 PM
iCloud
2:42 PM
Google
To: stellas1234@gmail.com
Details
Security alert
Today at 5:52 PM
Google
Wen Notes
New device signed in to
stellas1234@gmail.com
Your Google Account was just signed in to from a new Apple iPad device. You're getting this email to make sure it was you.
CHECK ACTIVITY
You received this email to let you know about important changes to your Google Account and services.
© 2018 Google Inc.,1600 Amphitheatre Parkway, Mountain View, CA 94043, USA

This is how your email page will look.

The column on the left shows emails in your inbox. Those marked with a blue dot have not yet been read.

Touch the email you wish to open.

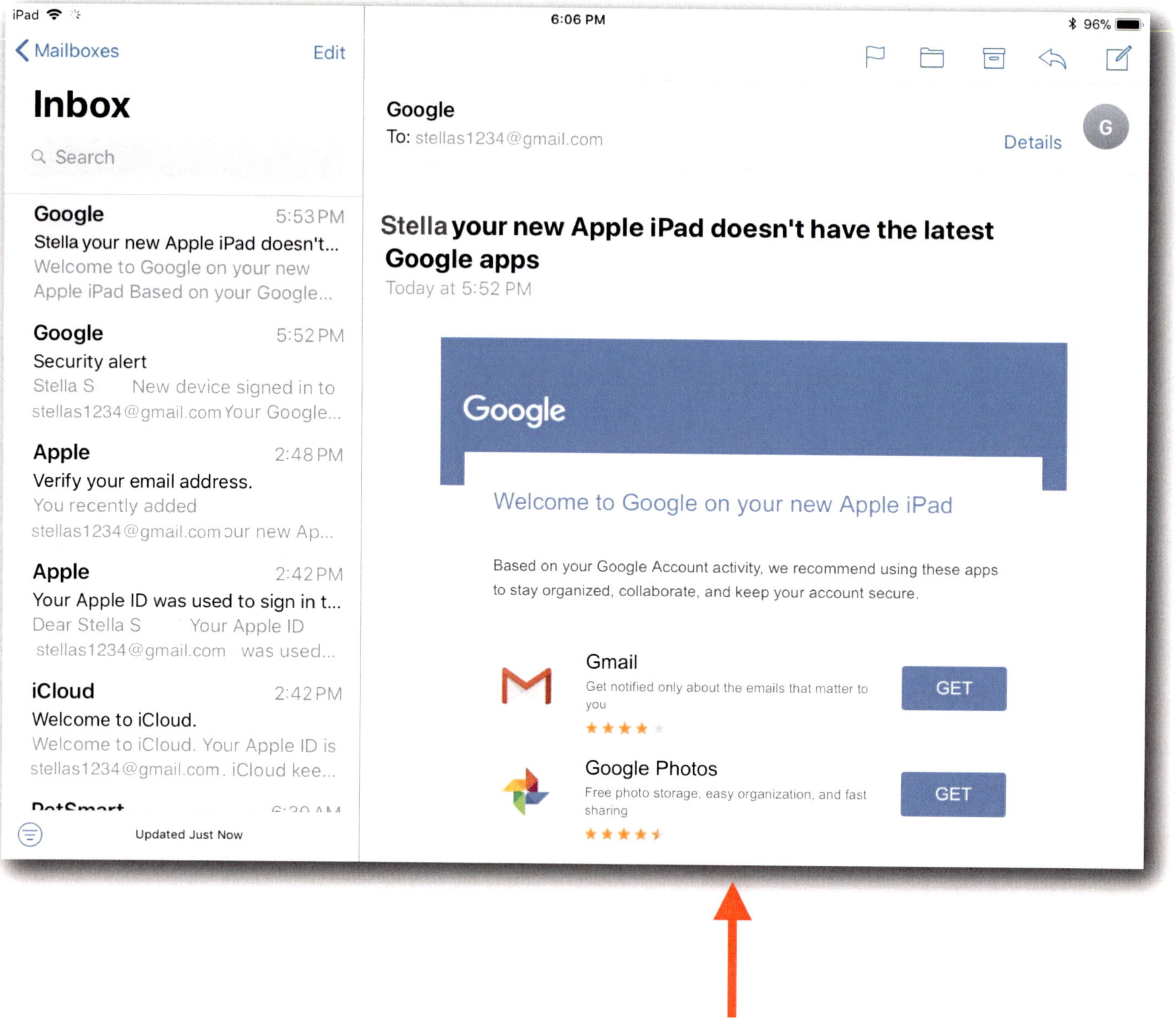
iPad
6:06 PM
96%
Mailboxes
Edit
Inbox
Search
Google 5:53 PM
Stella your new Apple iPad doesn't...
Welcome to Google on your new Apple iPad Based on your Google...
Google 5:52 PM
Security alert
Stella S New device signed in to stellas1234@gmail.com Your Google...
Apple 2:48 PM
Verify your email address.
You recently added stellas1234@gmail.com our new Ap...
Apple 2:42 PM
Your Apple ID was used to sign in t...
Dear Stella S Your Apple ID stellas1234@gmail.com was used...
iCloud 2:42 PM
Welcome to iCloud.
Welcome to iCloud. Your Apple ID is stellas1234@gmail.com. iCloud kee...
Updated Just Now
Google
To: stellas1234@gmail.com
Details
Stella your new Apple iPad doesn't have the latest Google apps
Today at 5:52 PM
Google
Welcome to Google on your new Apple iPad
Based on your Google Account activity, we recommend using these apps to stay organized, collaborate, and keep your account secure.
Gmail
Get notified only about the emails that matter to you
GET
Google Photos
Free photo storage, easy organization, and fast sharing
GET

The email you touch will open on the right side of the page.

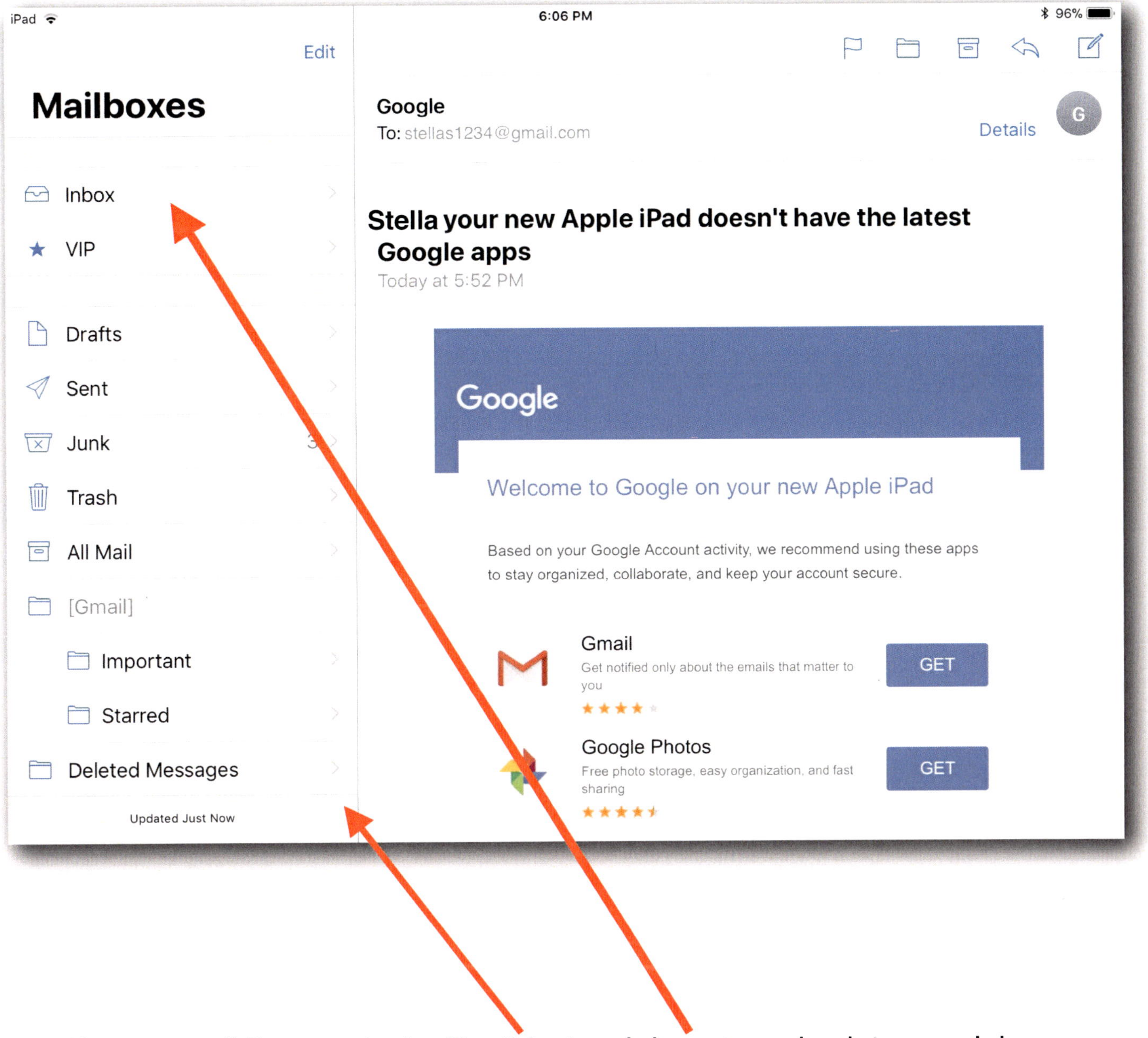
iPad
6:06 PM
96%
Edit
Mailboxes
Inbox
VIP
Drafts
Sent
Junk
Trash
All Mail
[Gmail]
Important
Starred
Deleted Messages
Updated Just Now
Google
To: stellas1234@gmail.com
Details
Stella your new Apple iPad doesn't have the latest Google apps
Today at 5:52 PM
Google
Welcome to Google on your new Apple iPad
Based on your Google Account activity, we recommend using these apps to stay organized, collaborate, and keep your account secure.
Gmail
Get notified only about the emails that matter to you
GET
Google Photos
Free photo storage, easy organization, and fast sharing
GET

If your email list ever looks like this, touch here to go back to your inbox.

Press the Home Button to close.

Adding Contacts
(Address Book)

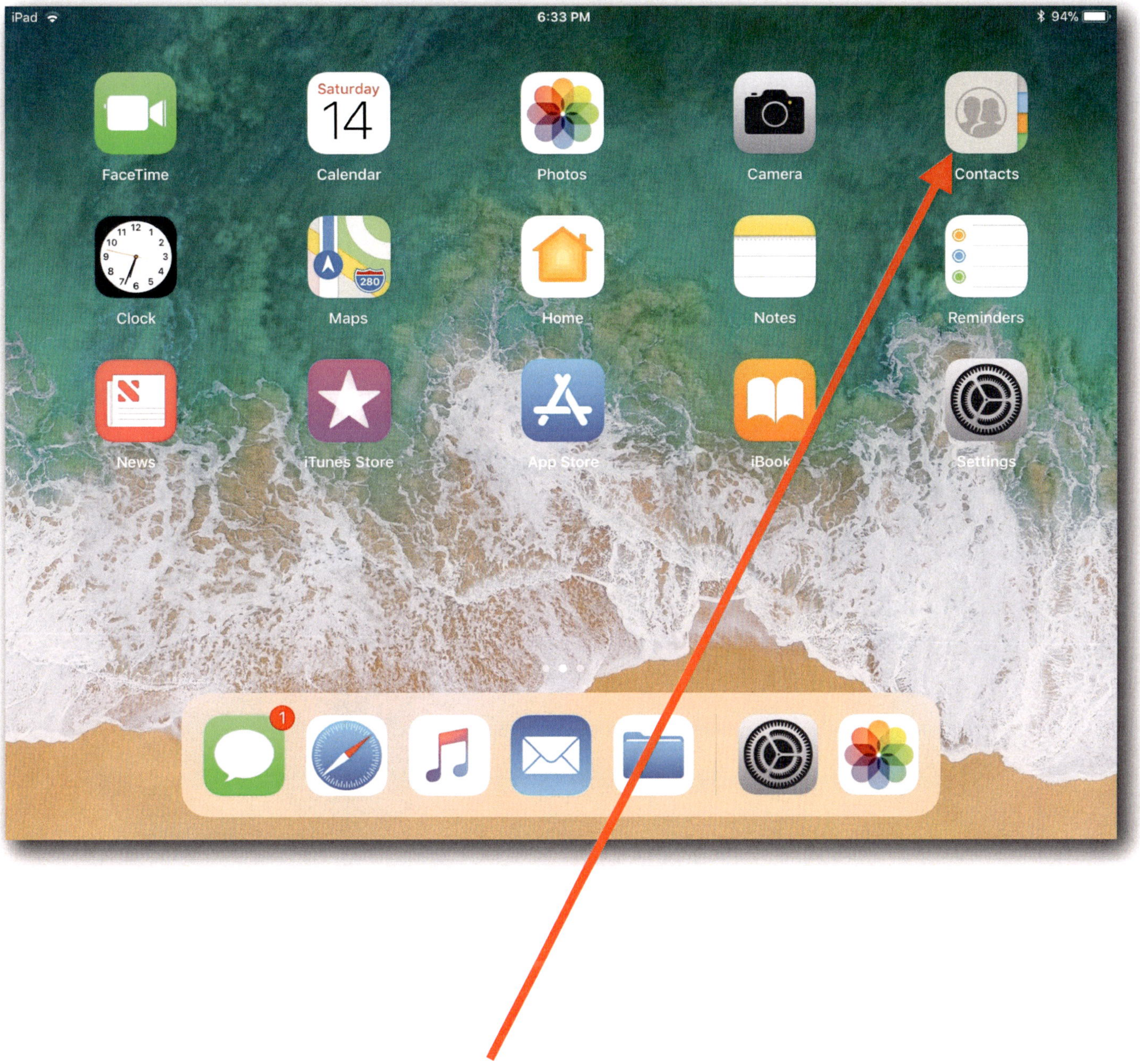

Touch here to open Contacts.

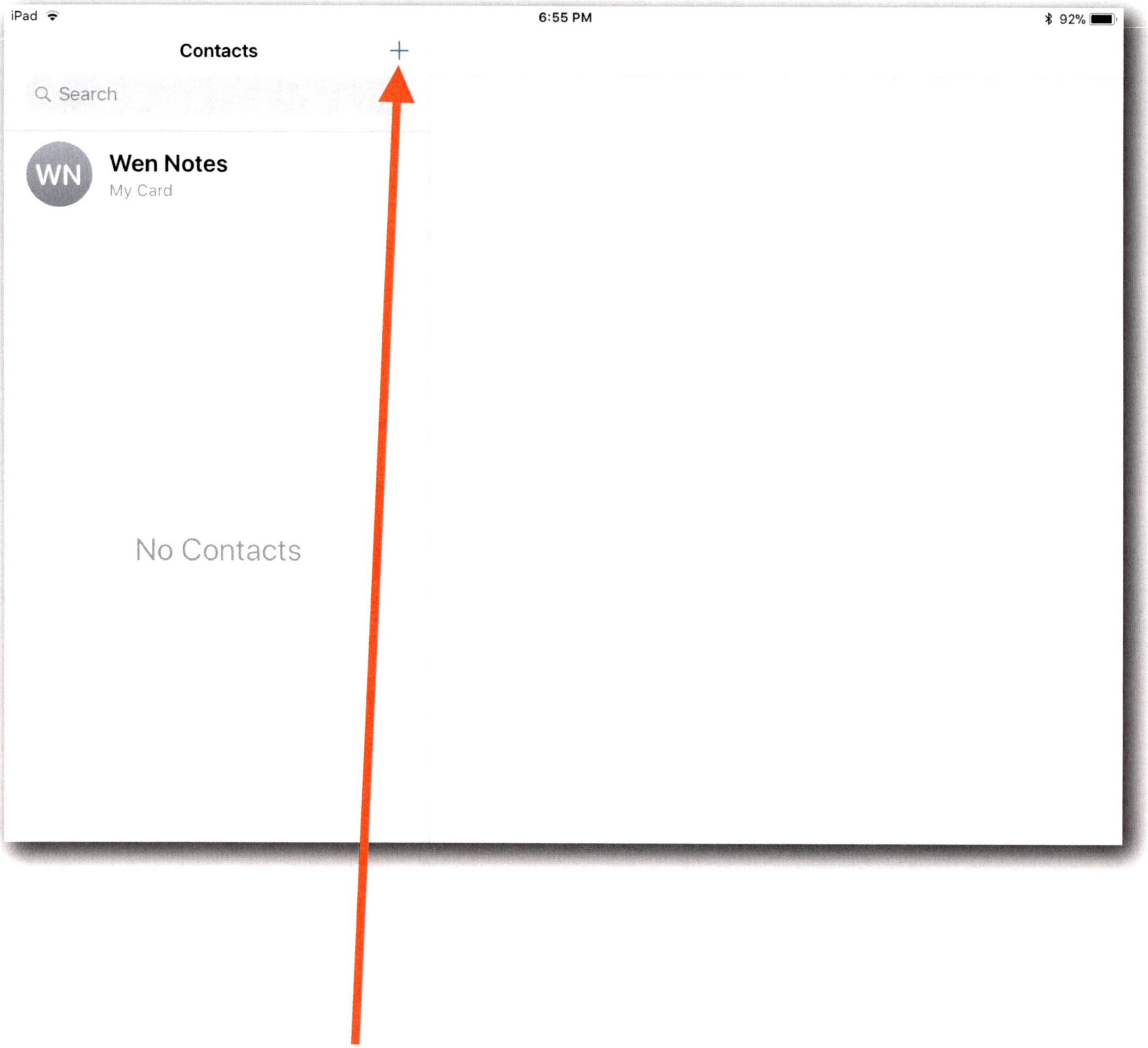

Touch here to add a contact to your Contacts File.

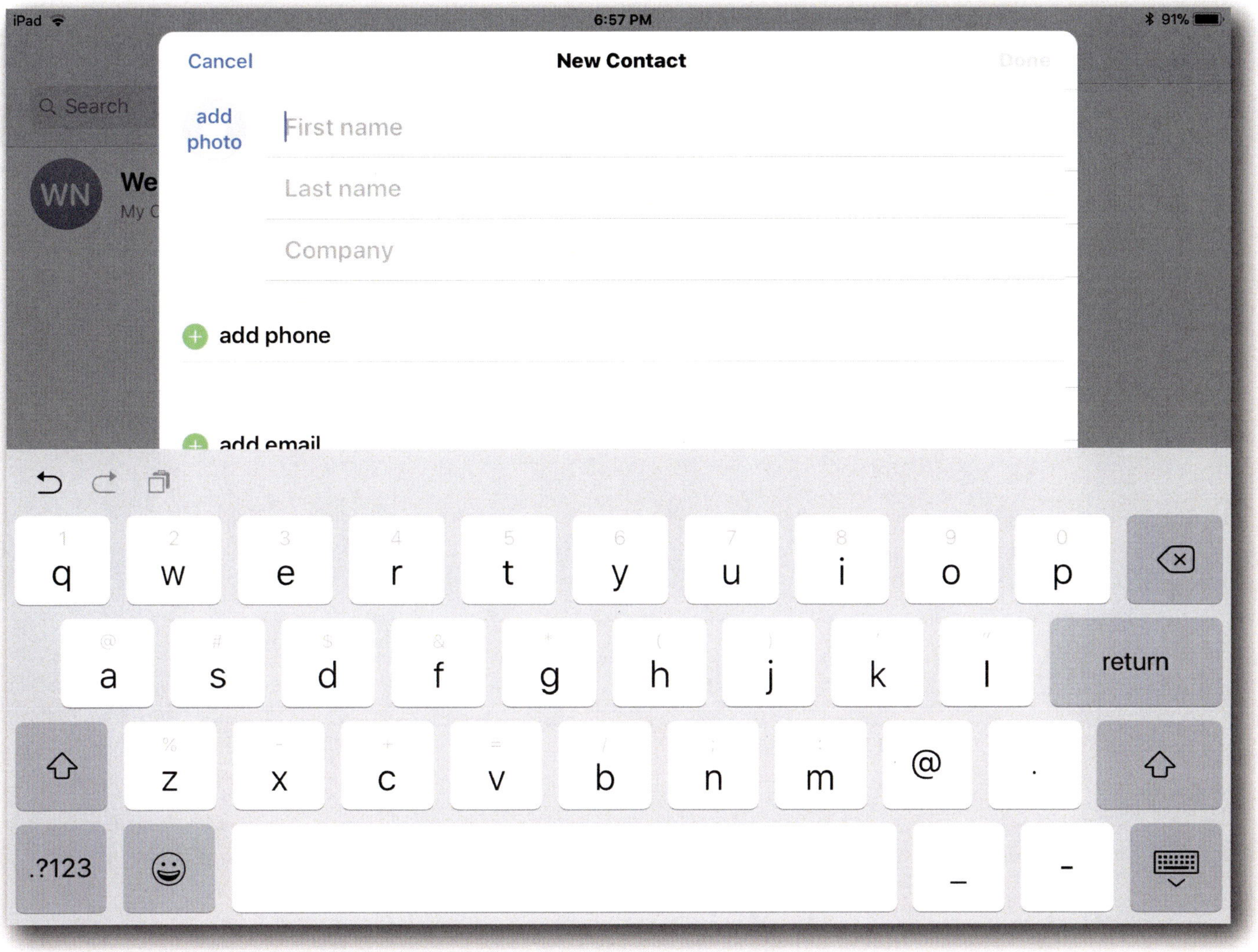

Type in the first name of your new contact.

Then touch each field you would like to fill in, such as:
Last name, add phone, add email.

Note: You may need to scroll down (slide finger up on the screen) in order to touch fields which may be hidden by the keyboard.

Not all fields need to be filled in.

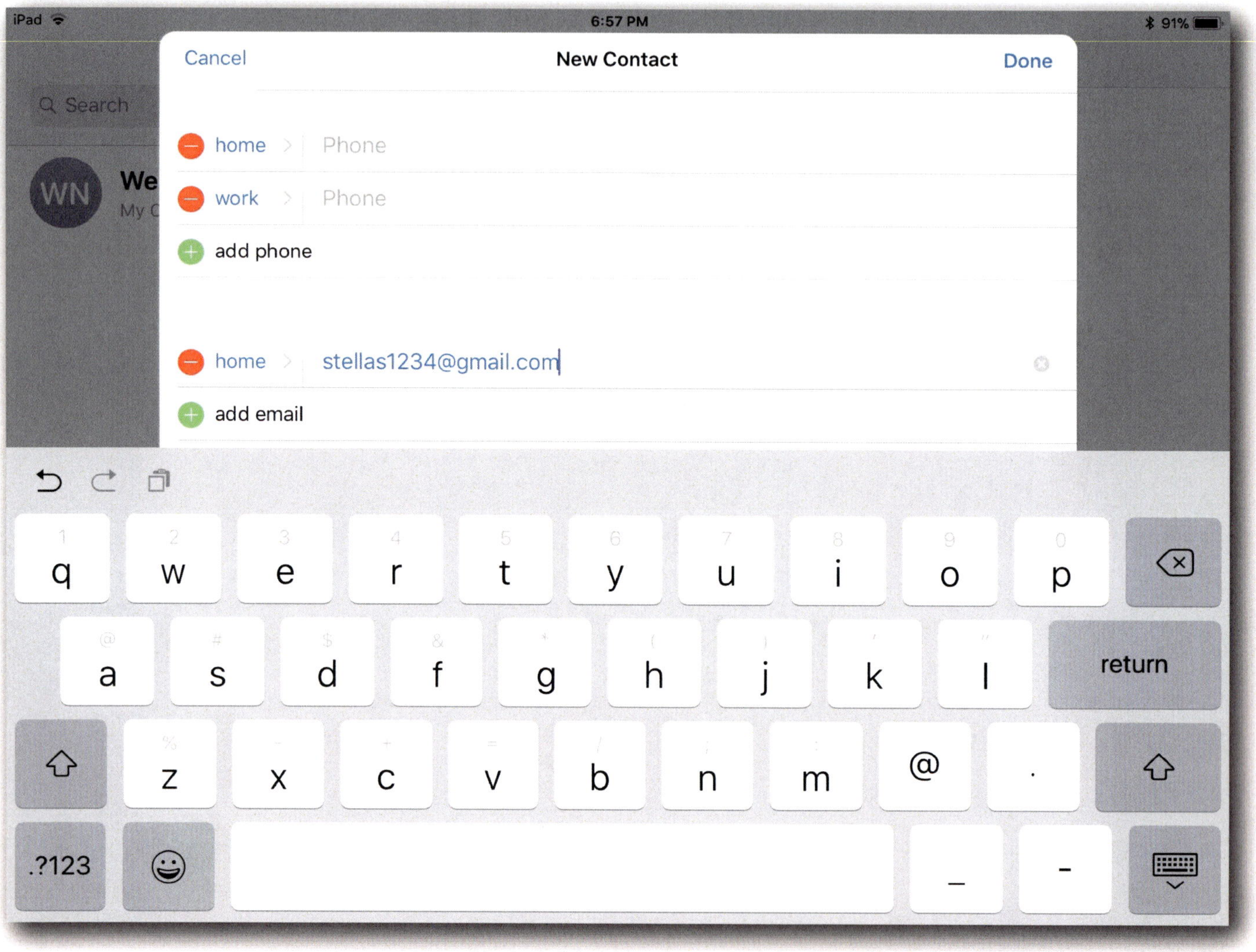

After touching a field, type in the information.
Then touch the next field you wish to fill in and type in the information.

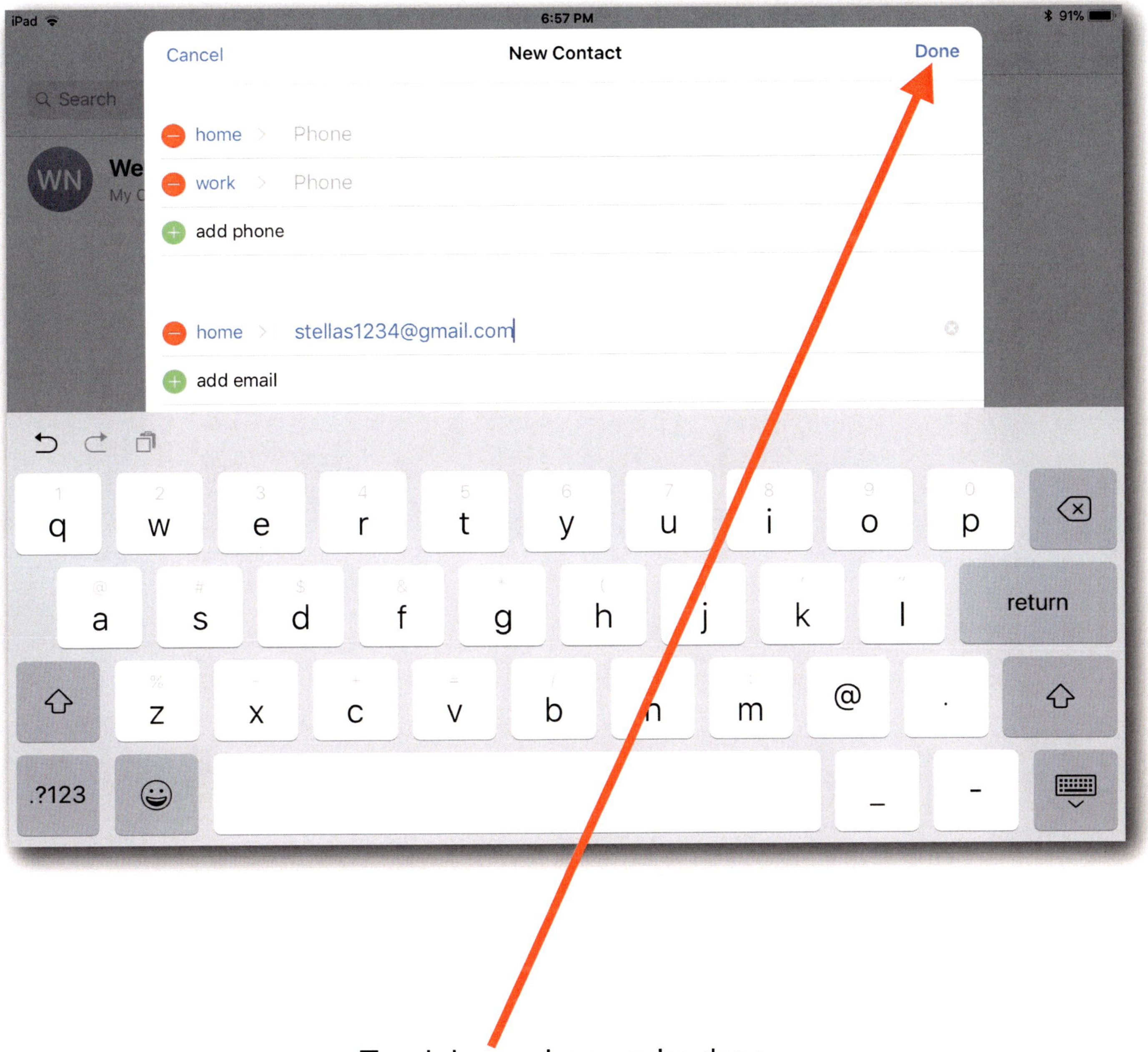

Touch here when you're done.

Repeat the process to add others into Contacts.
Press the Home Button when you're done.

Using The Internet

Touch here to open "Safari." This is the internet app.

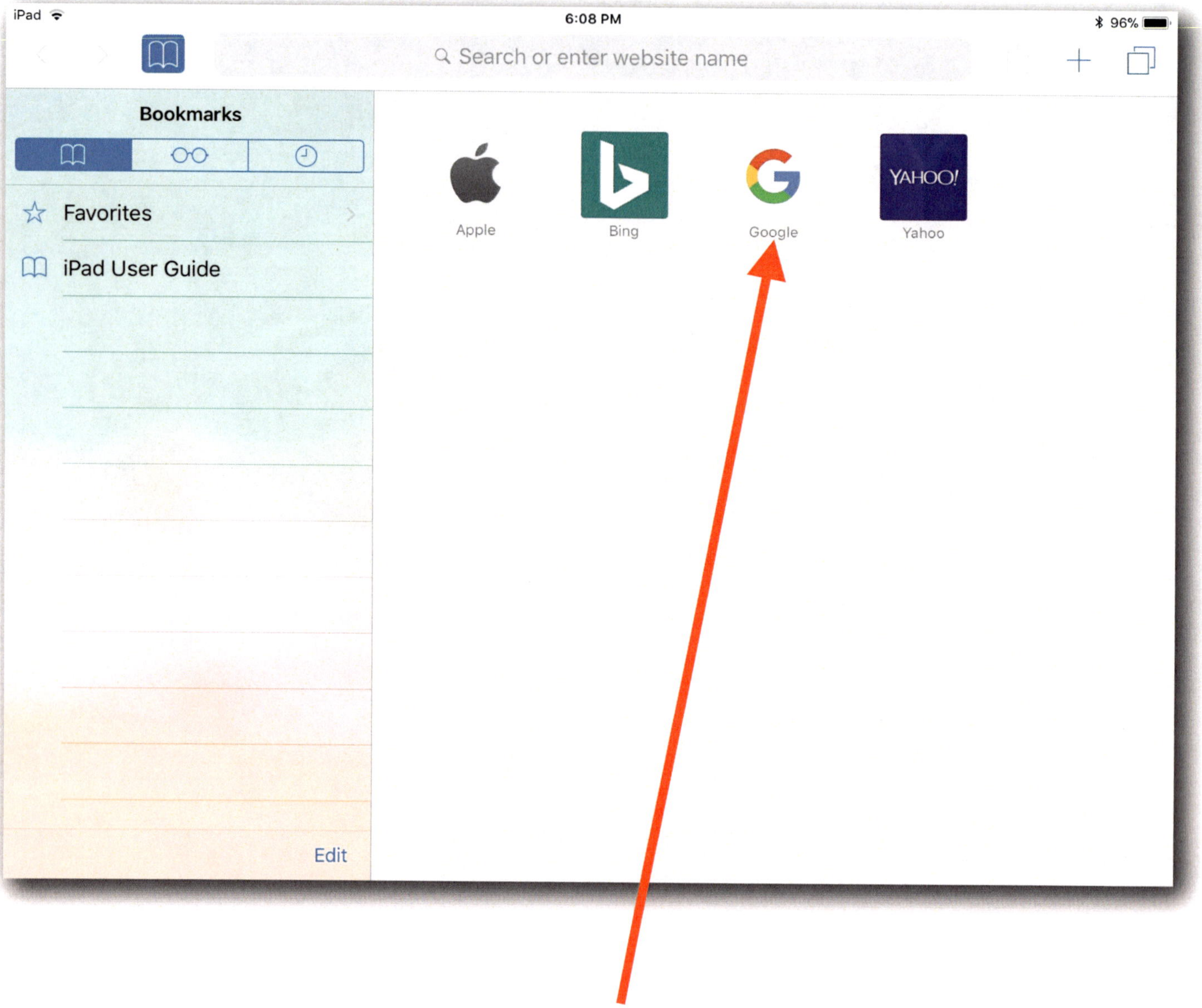

Touch the internet browser you'd like to use on your iPad.

For this step I will be using Google to demonstrate.

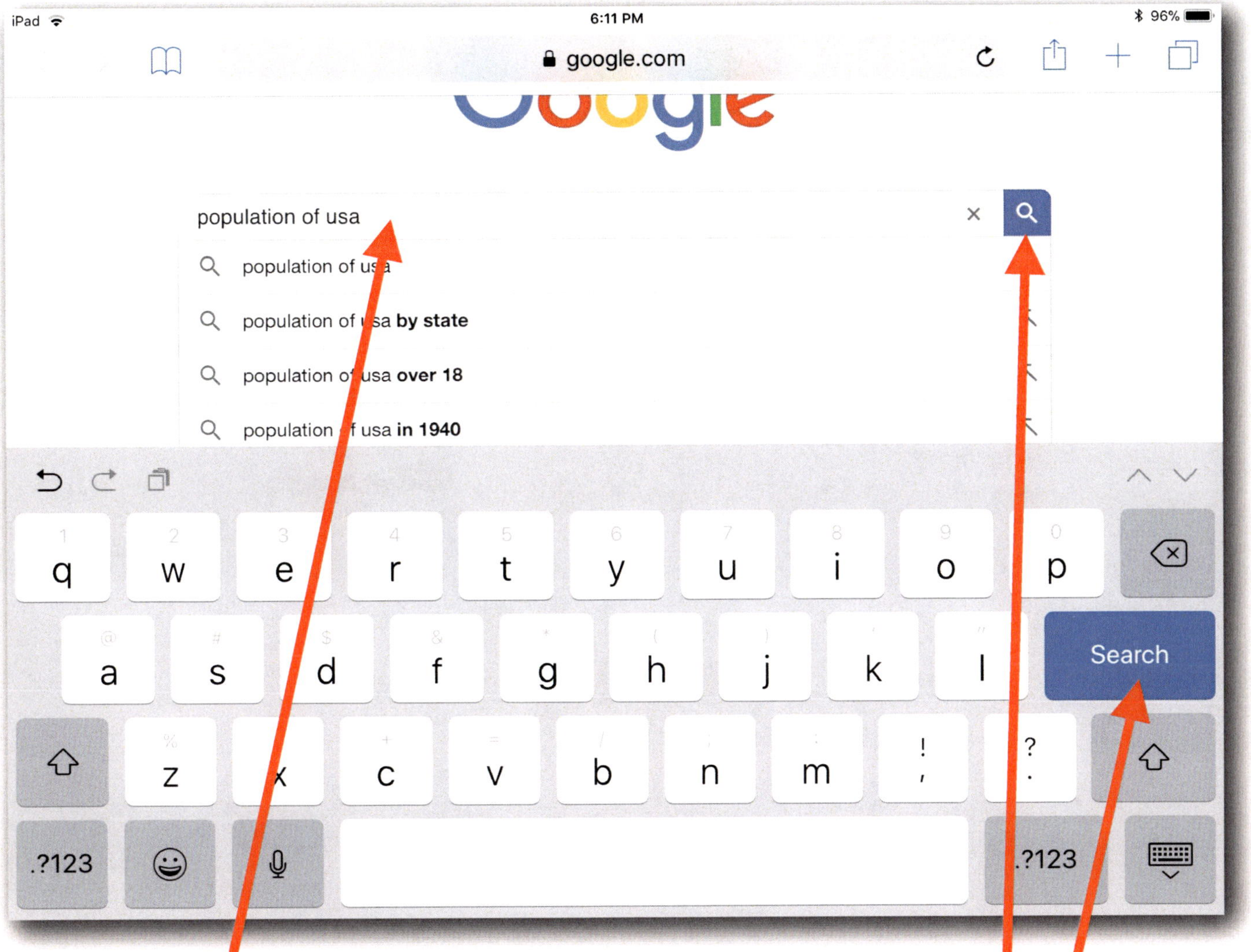

Touch here and type what you'd like to look up.

Touch either one of these
to initiate your search.

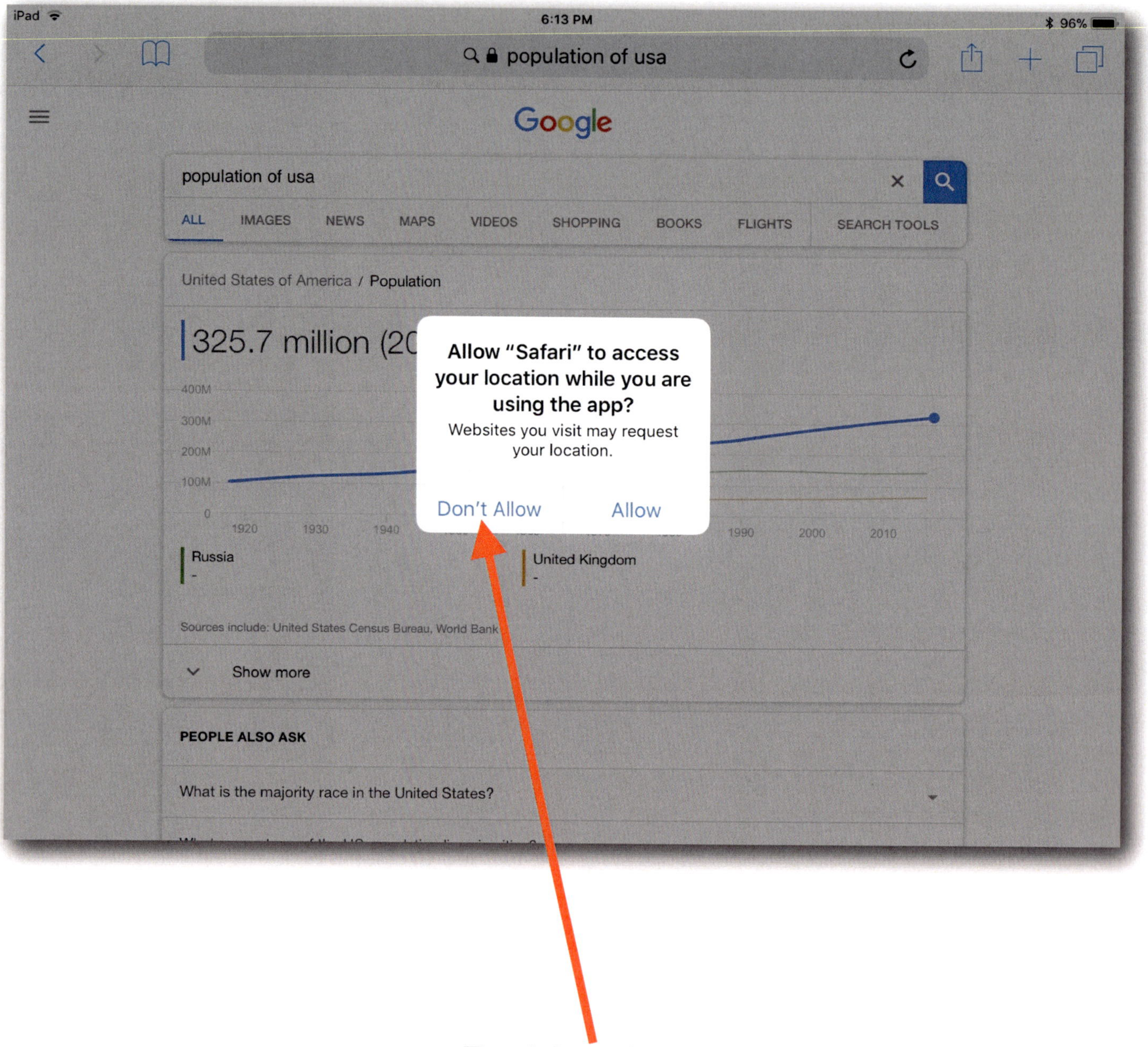

Touch here for now.

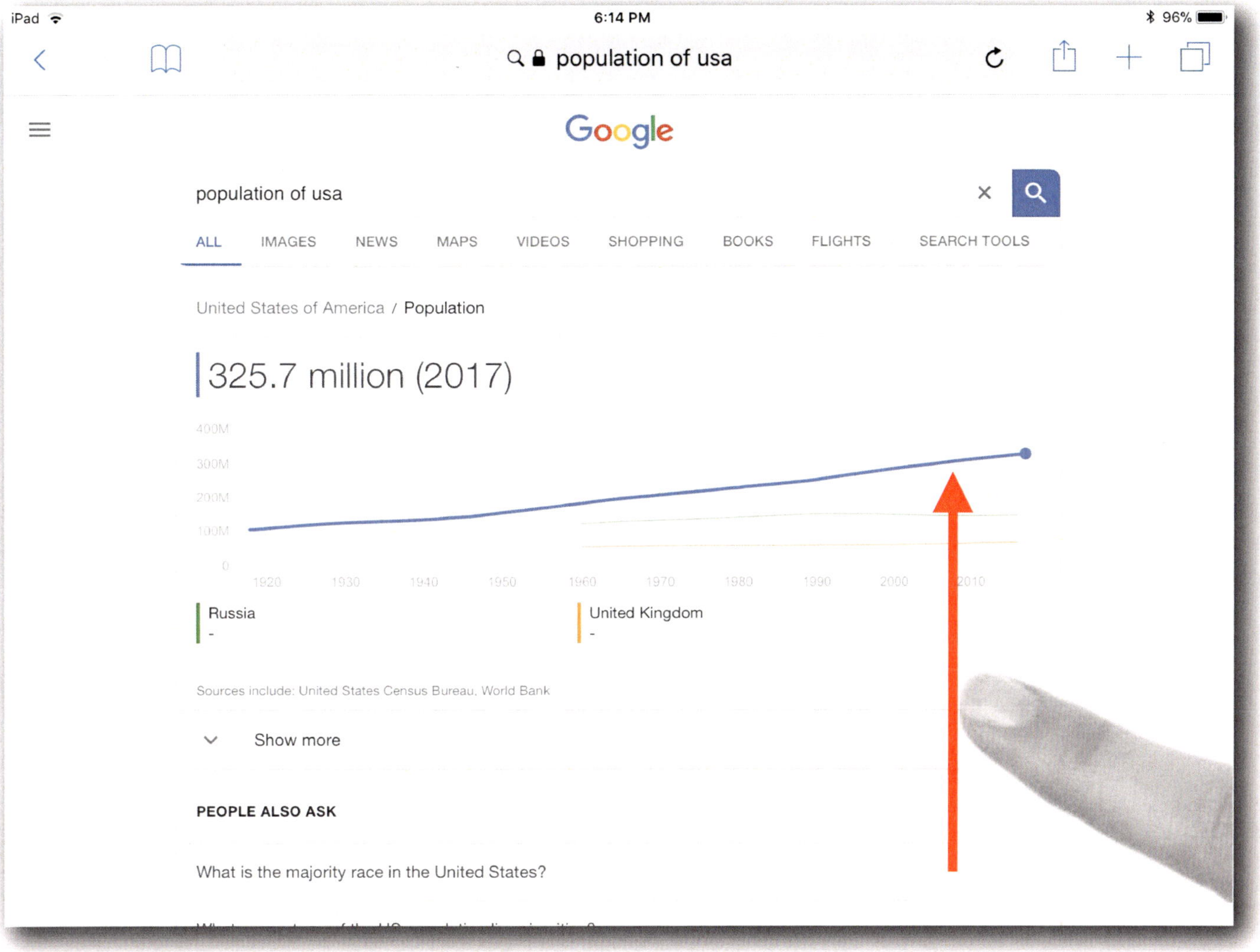

Several search results will appear.
You can scroll down until you find one you'd like to open.

Note: Scroll down the page by gently sliding your finger up the on the screen.

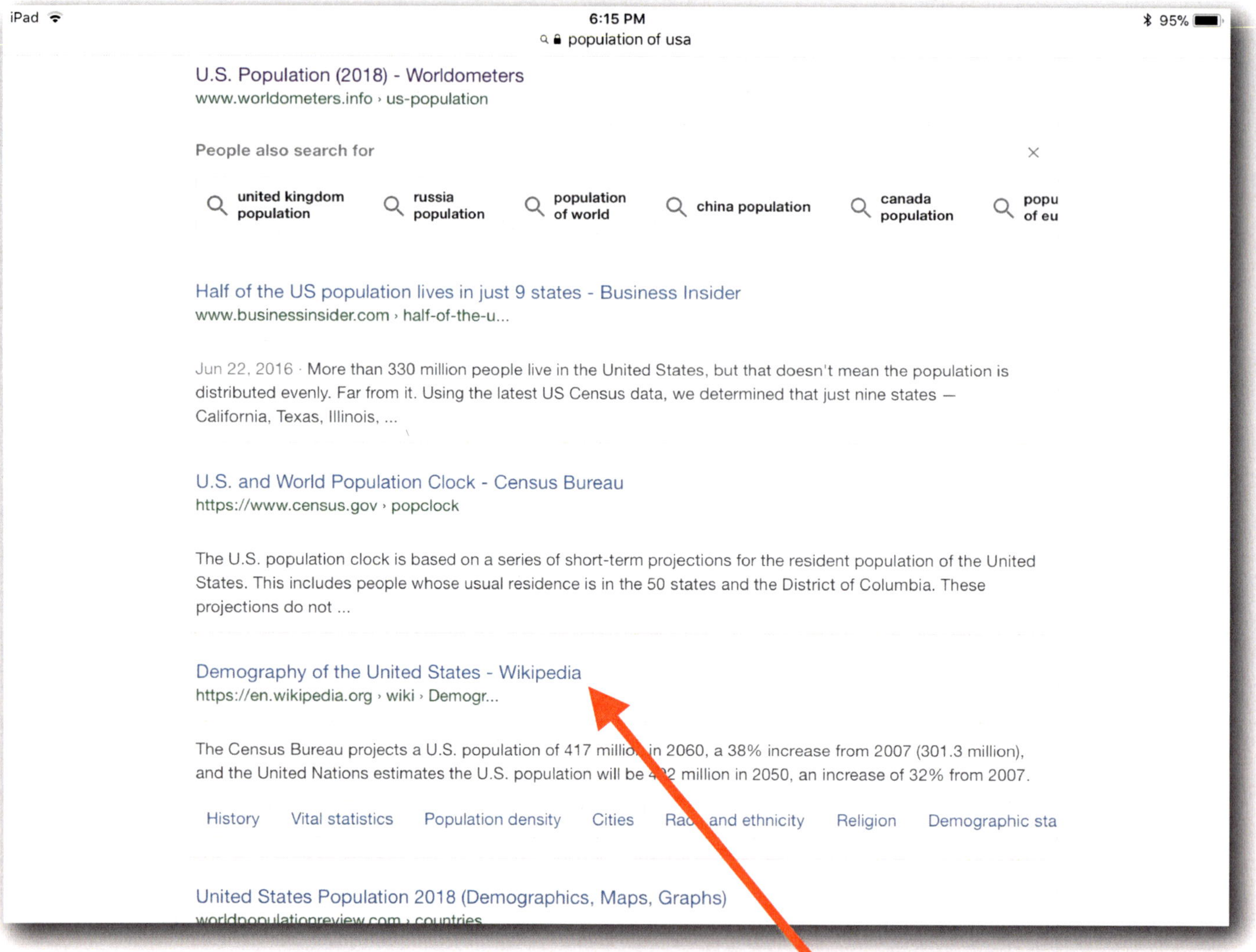
iPad
6:15 PM
population of usa
95%
U.S. Population (2018) - Worldometers
www.worldometers.info › us-population
People also search for
united kingdom population
russia population
population of world
china population
canada population
popu of eu
Half of the US population lives in just 9 states - Business Insider
www.businessinsider.com › half-of-the-u...
Jun 22, 2016 · More than 330 million people live in the United States, but that doesn't mean the population is distributed evenly. Far from it. Using the latest US Census data, we determined that just nine states — California, Texas, Illinois, ...
U.S. and World Population Clock - Census Bureau
https://www.census.gov › popclock
The U.S. population clock is based on a series of short-term projections for the resident population of the United States. This includes people whose usual residence is in the 50 states and the District of Columbia. These projections do not ...
Demography of the United States - Wikipedia
https://en.wikipedia.org › wiki › Demogr...
History
Vital statistics
Population density
Cities
Religion
Demographic sta
United States Population 2018 (Demographics, Maps, Graphs)

For the purpose of this demonstration, I'm touching this one.

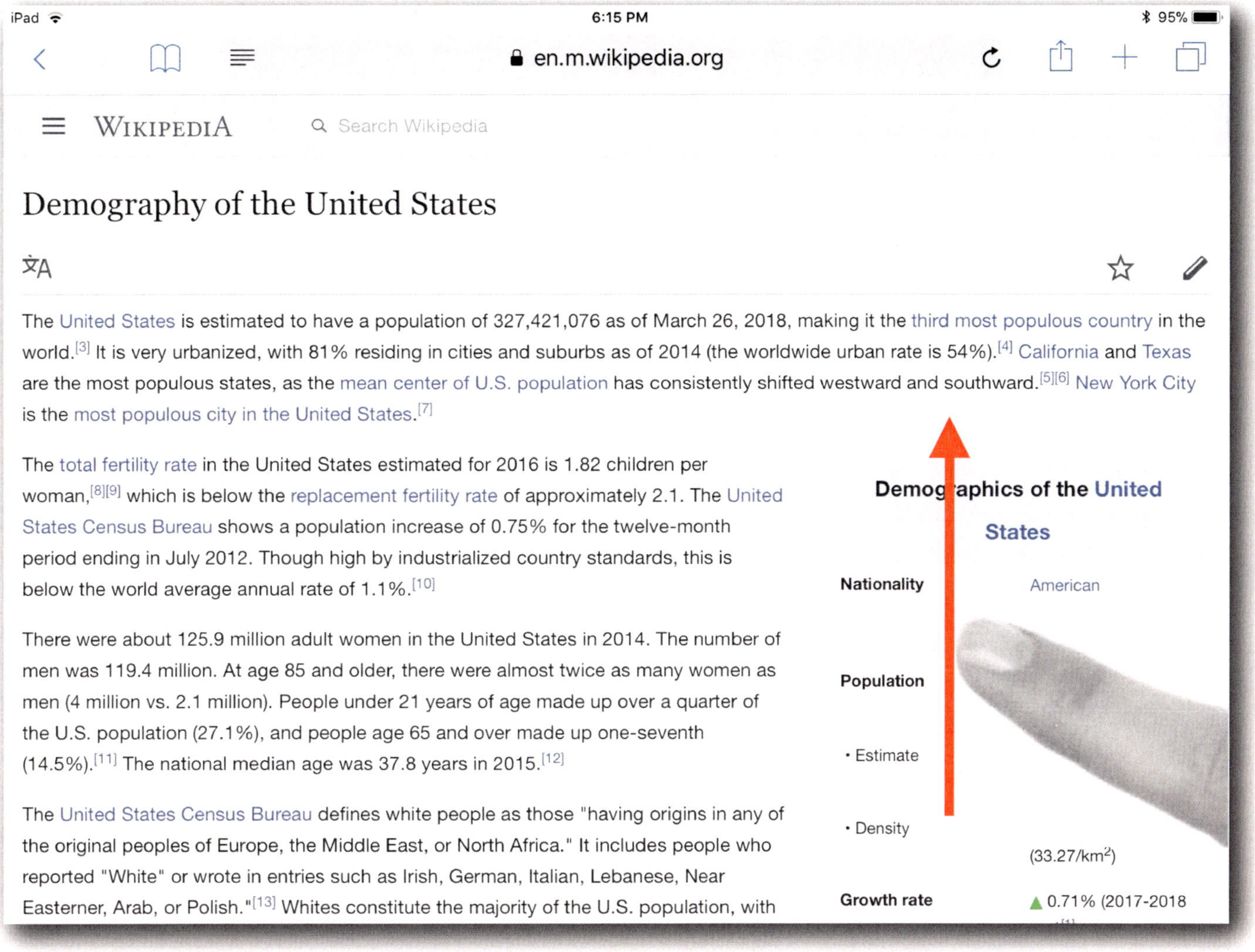
iPad
6:15 PM
95%
en.m.wikipedia.org
Wikipedia
Search Wikipedia
Demography of the United States
The United States is estimated to have a population of 327,421,076 as of March 26, 2018, making it the third most populous country in the world.[3] It is very urbanized, with 81% residing in cities and suburbs as of 2014 (the worldwide urban rate is 54%).[4] California and Texas are the most populous states, as the mean center of U.S. population has consistently shifted westward and southward.[5][6] New York City is the most populous city in the United States.[7]
The total fertility rate in the United States estimated for 2016 is 1.82 children per woman,[8][9] which is below the replacement fertility rate of approximately 2.1. The United States Census Bureau shows a population increase of 0.75% for the twelve-month period ending in July 2012. Though high by industrialized country standards, this is below the world average annual rate of 1.1%.[10]
There were about 125.9 million adult women in the United States in 2014. The number of men was 119.4 million. At age 85 and older, there were almost twice as many women as men (4 million vs. 2.1 million). People under 21 years of age made up over a quarter of the U.S. population (27.1%), and people age 65 and over made up one-seventh (14.5%).[11] The national median age was 37.8 years in 2015.[12]
The United States Census Bureau defines white people as those "having origins in any of the original peoples of Europe, the Middle East, or North Africa." It includes people who reported "White" or wrote in entries such as Irish, German, Italian, Lebanese, Near Easterner, Arab, or Polish."[13] Whites constitute the majority of the U.S. population, with
Demographics of the United States
Nationality
American
Population
• Estimate
• Density
(33.27/km²)
Growth rate
0.71% (2017-2018

You can scroll down to read more of this page,
again by gently sliding your finger up the on the screen.

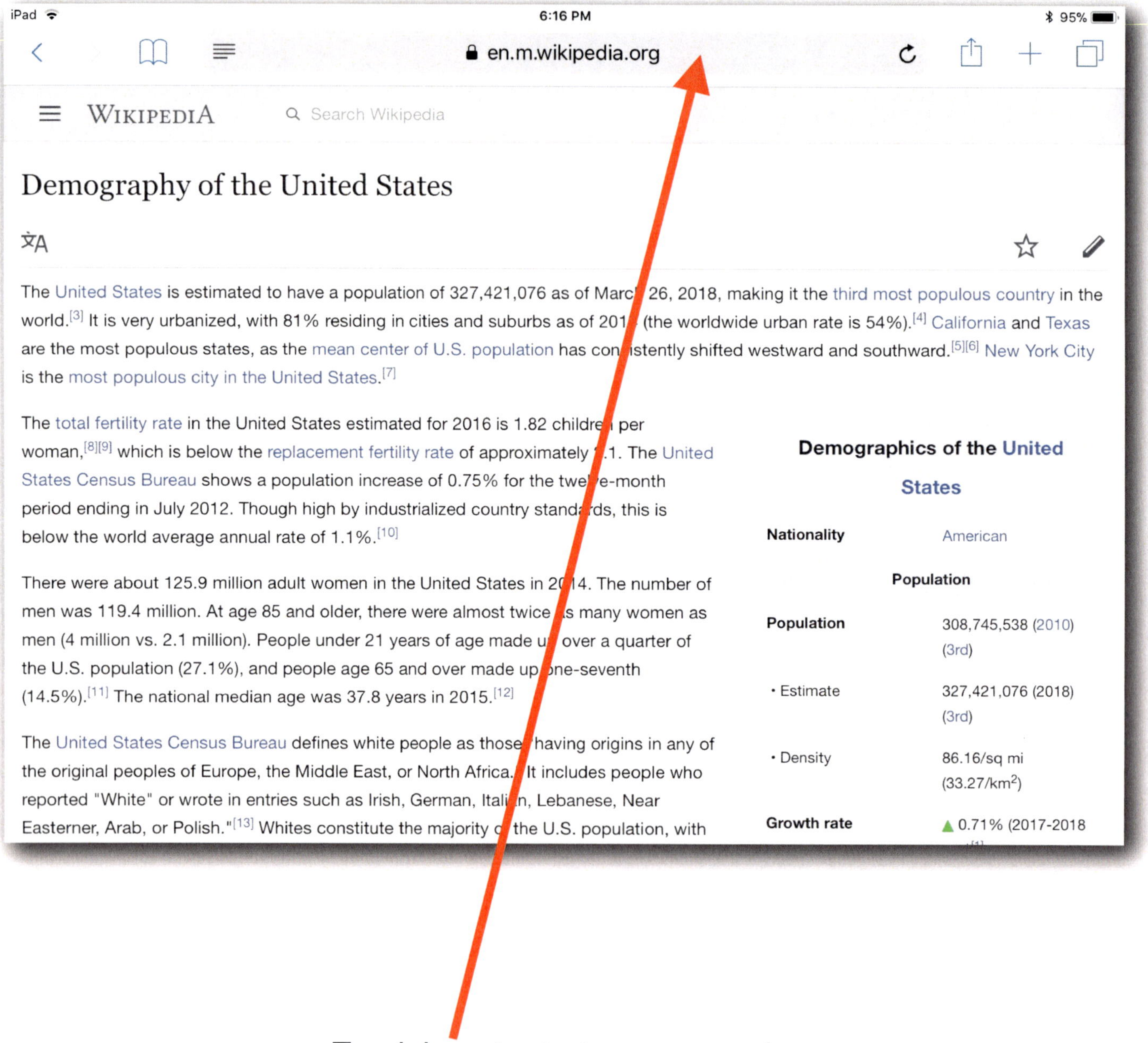
iPad
6:16 PM
95%
en.m.wikipedia.org
Wikipedia
Search Wikipedia
Demography of the United States
The United States is estimated to have a population of 327,421,076 as of March 26, 2018, making it the third most populous country in the world.[3] It is very urbanized, with 81% residing in cities and suburbs as of 2014 (the worldwide urban rate is 54%).[4] California and Texas are the most populous states, as the mean center of U.S. population has consistently shifted westward and southward.[5][6] New York City is the most populous city in the United States.[7]
Demographics of the United States
Nationality
American
Population
Population
308,745,538 (2010) (3rd)
• Estimate
327,421,076 (2018) (3rd)
• Density
86.16/sq mi (33.27/km2)
Growth rate
0.71% (2017-2018

Touch here to start a new search.

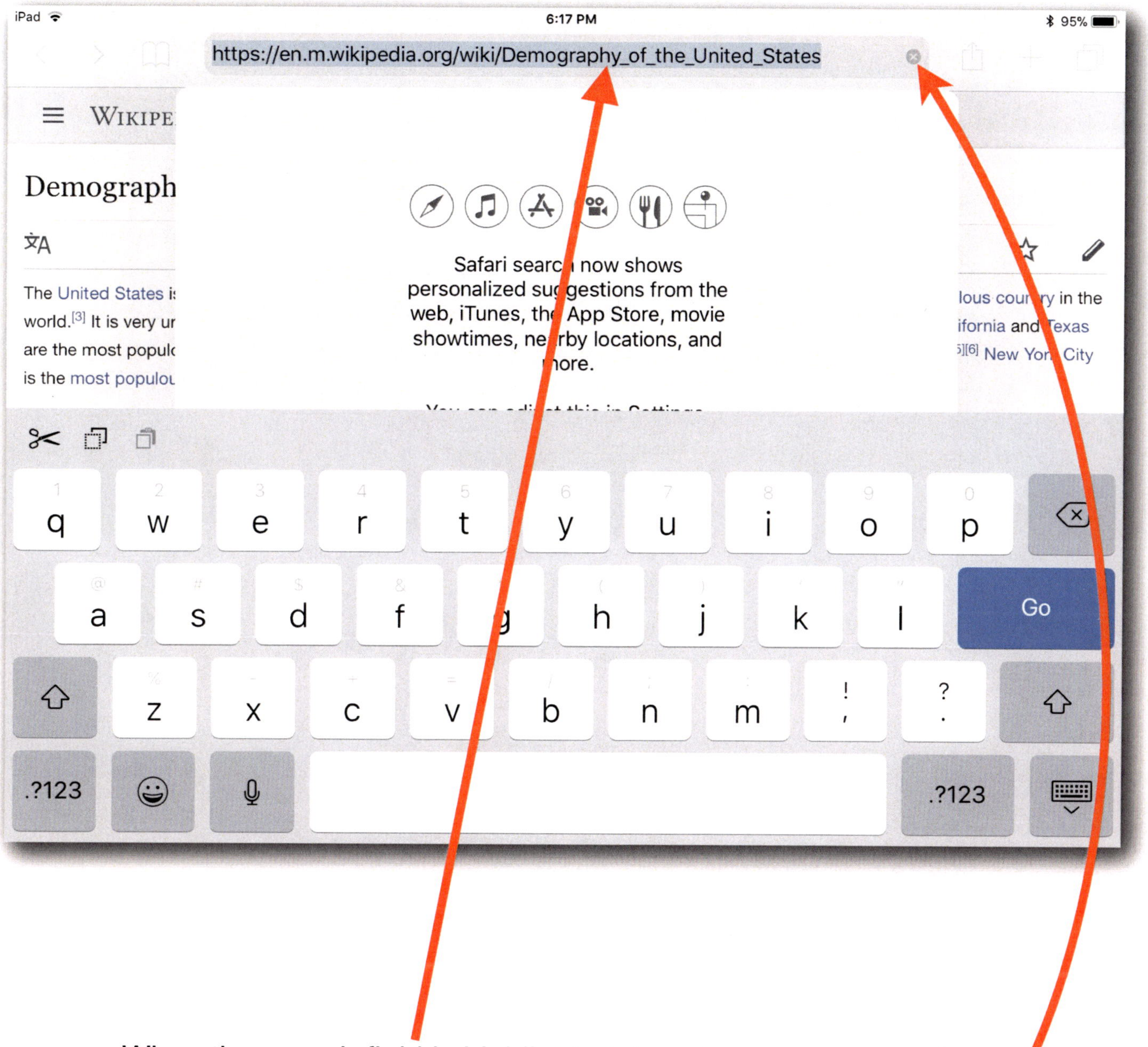
iPad
6:17 PM
95%
https://en.m.wikipedia.org/wiki/Demography_of_the_United_States
Safari search now shows personalized suggestions from the web, iTunes, the App Store, movie showtimes, nearby locations, and more.
Go
.?123

When the search field is highlighted, type in your new search.

The highlighted text should disappear when you start typing in your new search. If it doesn't touch here.

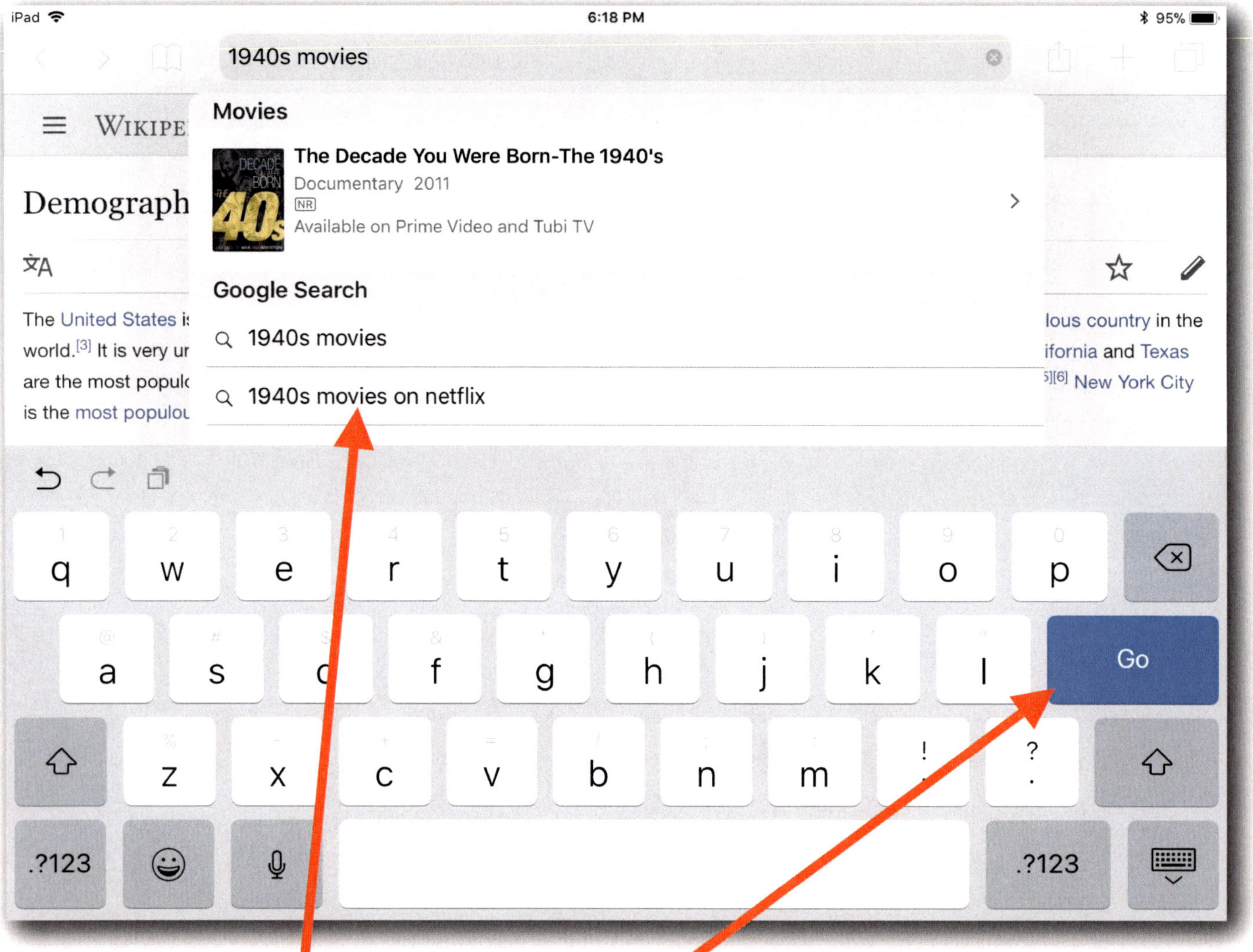

Touch here to initiate your new search.

OR…

Touch one of the suggestions from the “drop down” menu.

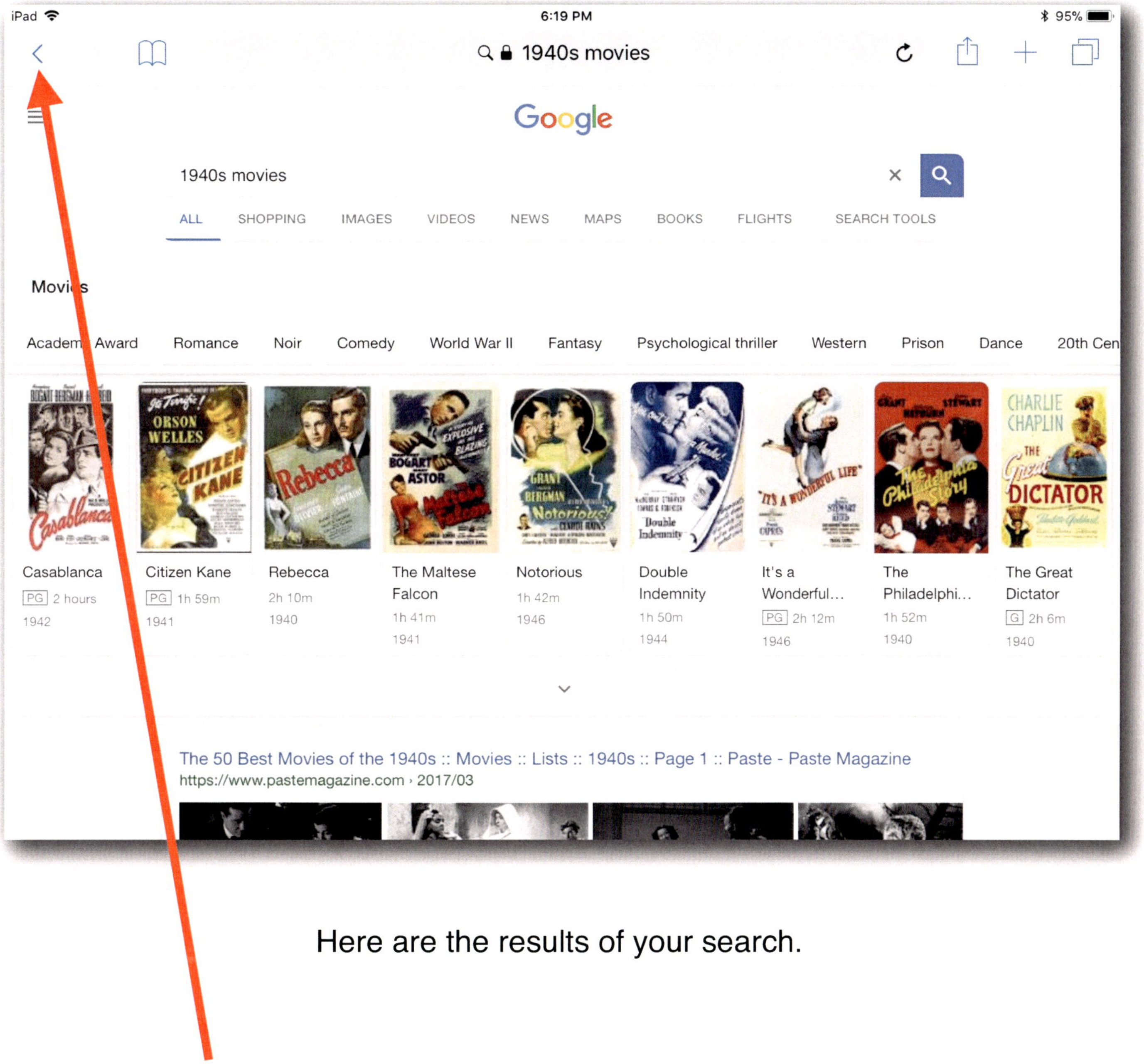

Here are the results of your search.

Touch here if you choose to go back to the previous internet page.

Press the Home Button when you're done.

Setting Up Text Messaging

Touch here to open Text Messaging.

The first time you use Text Messaging you will need to set it up.

Touch here to enter your Apple ID password.

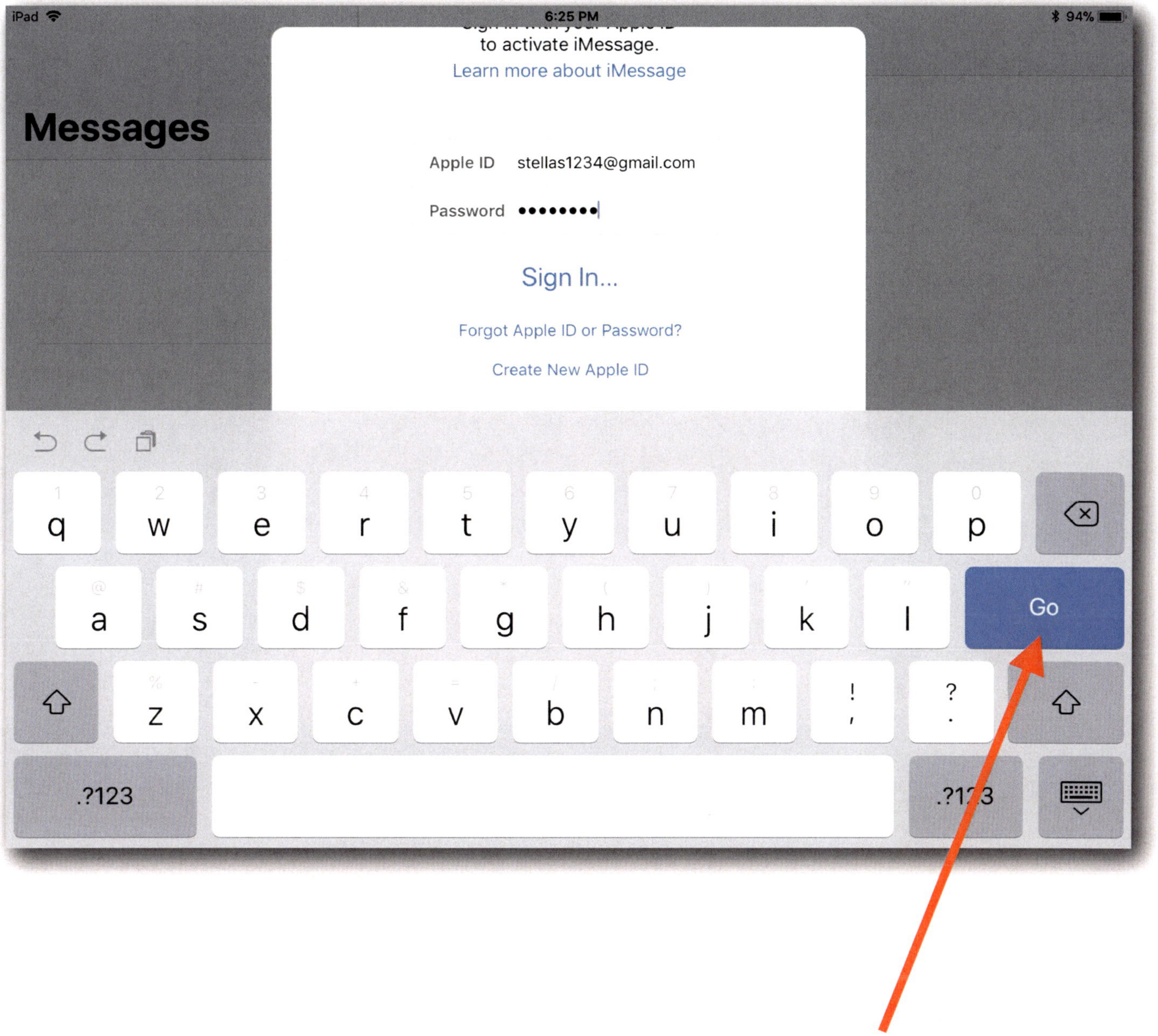

Type in your Apple ID password. Then touch here.

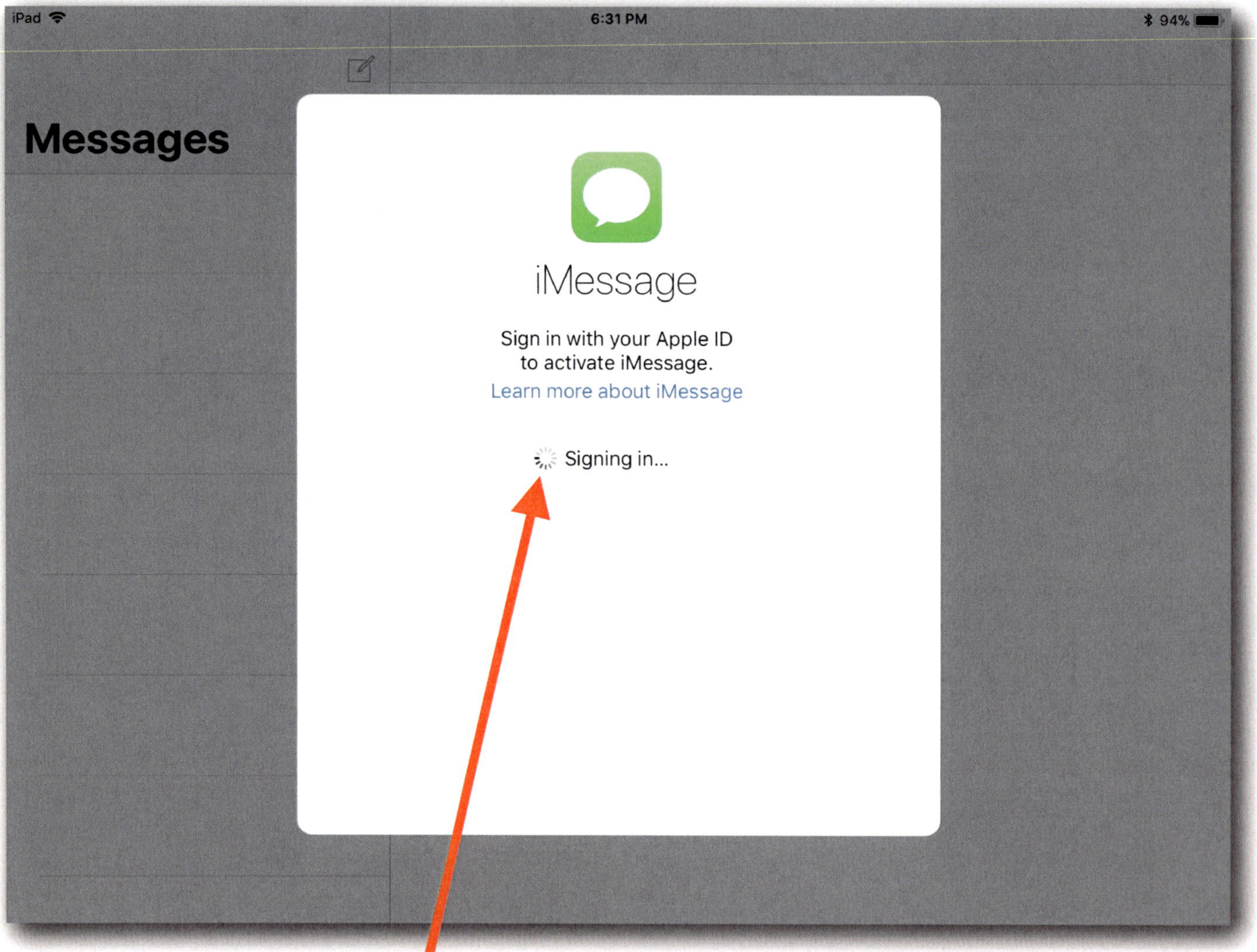

The spinning wheel shows that your iPad is signing you in.

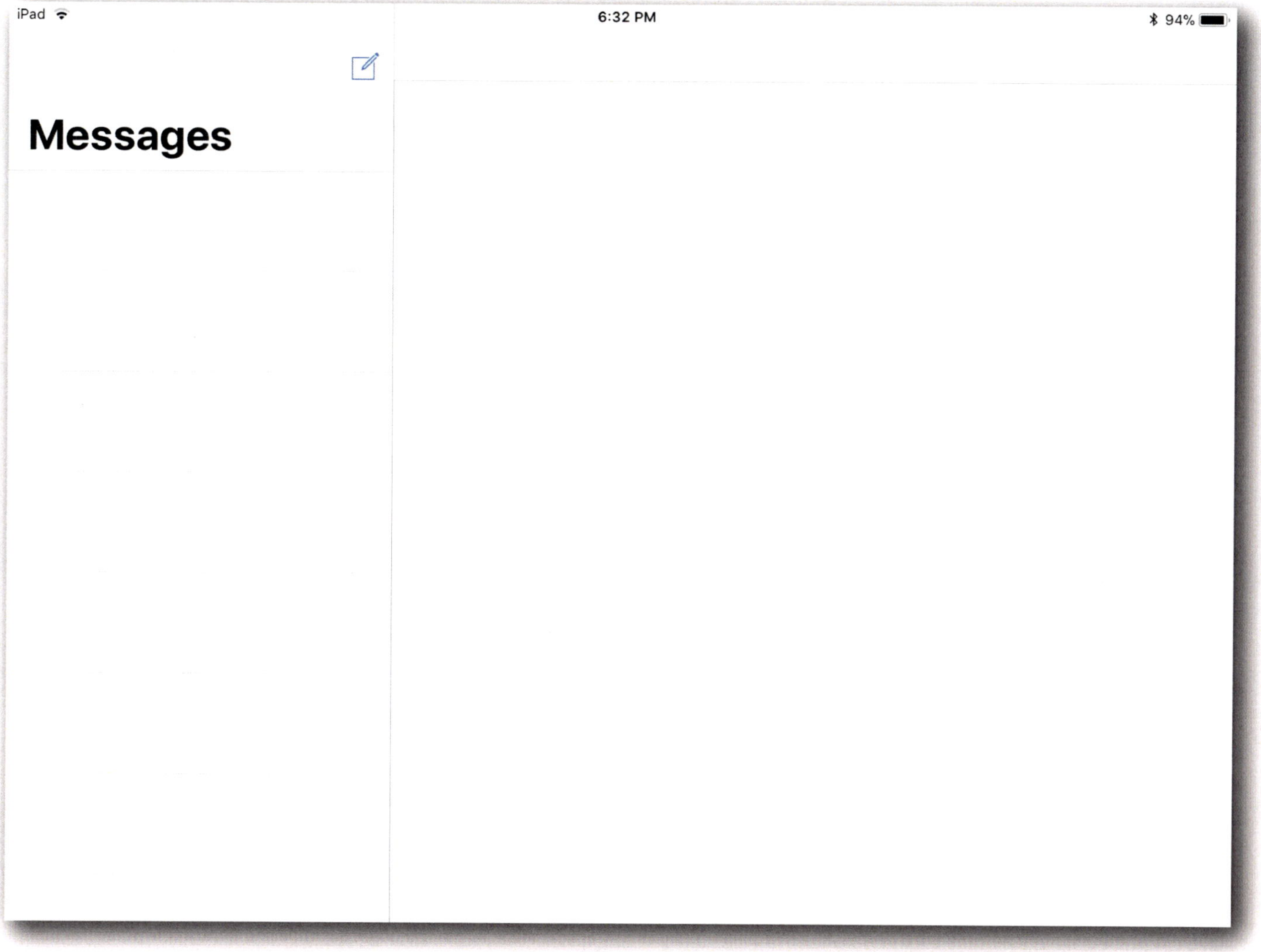

Welcome to Text Messaging!

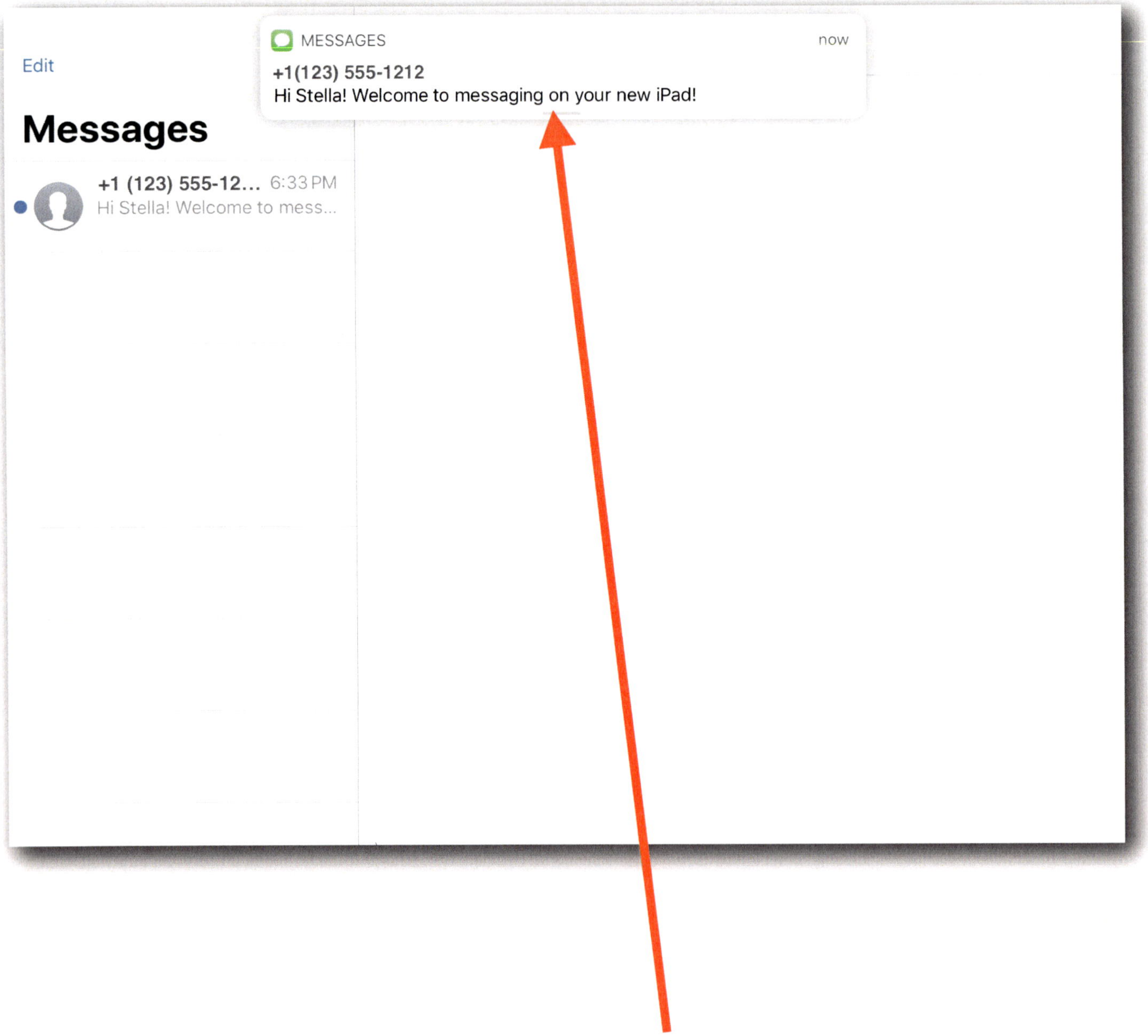
Edit
Messages
MESSAGES
now
+1(123) 555-1212
Hi Stella! Welcome to messaging on your new iPad!
+1 (123) 555-12...
6:33 PM
Hi Stella! Welcome to mess...

Your iPad will alert you with a notification like this when a new message comes in.

You may swipe the notification up to get rid of it, touch the notification to open the message, or just wait and it will go away on its own.

Using Text Messaging

This number indicates that you have new messages.

Touch the icon to open text messaging.

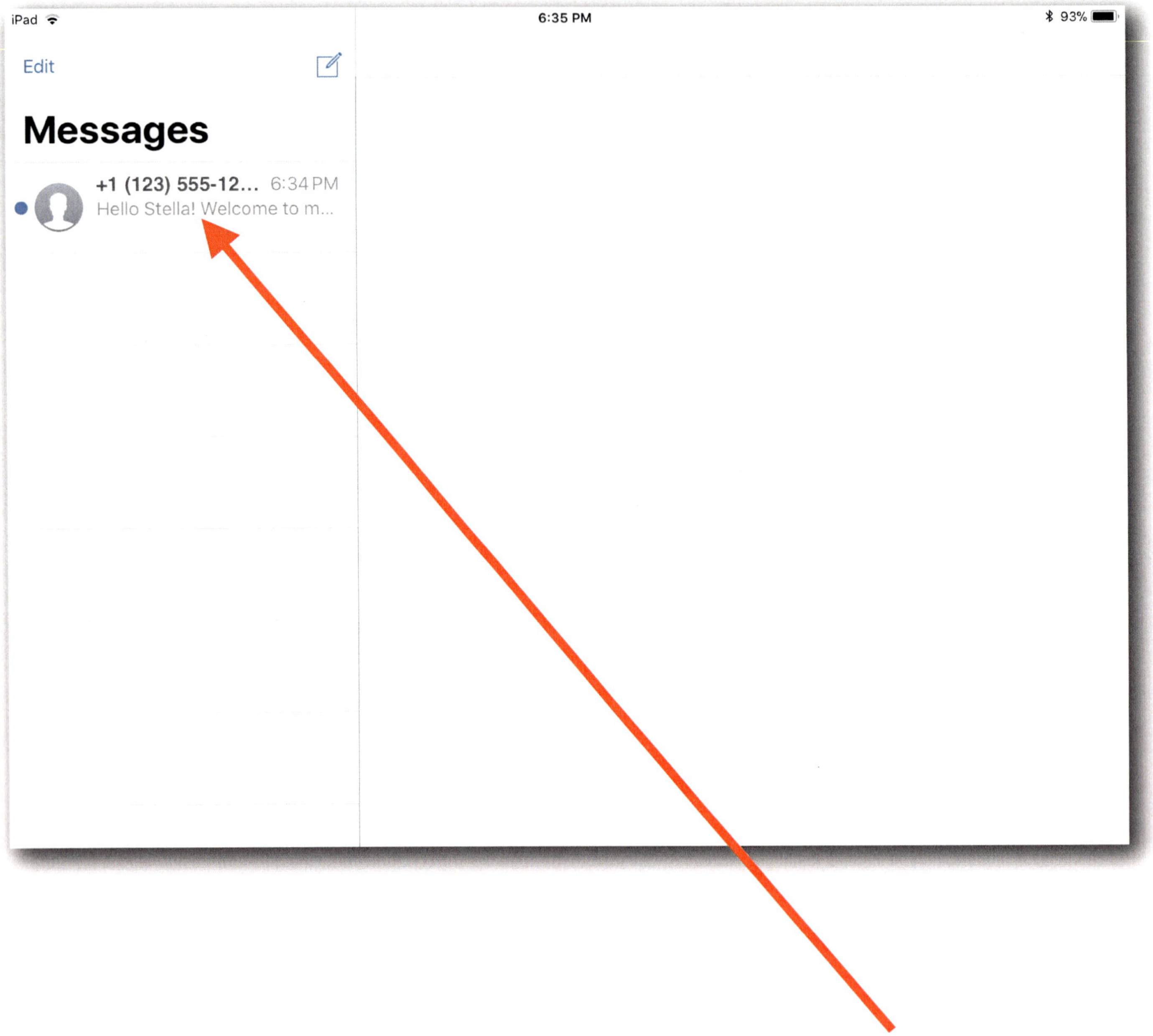

The blue dot indicates an unread message. Touch here to open the message.

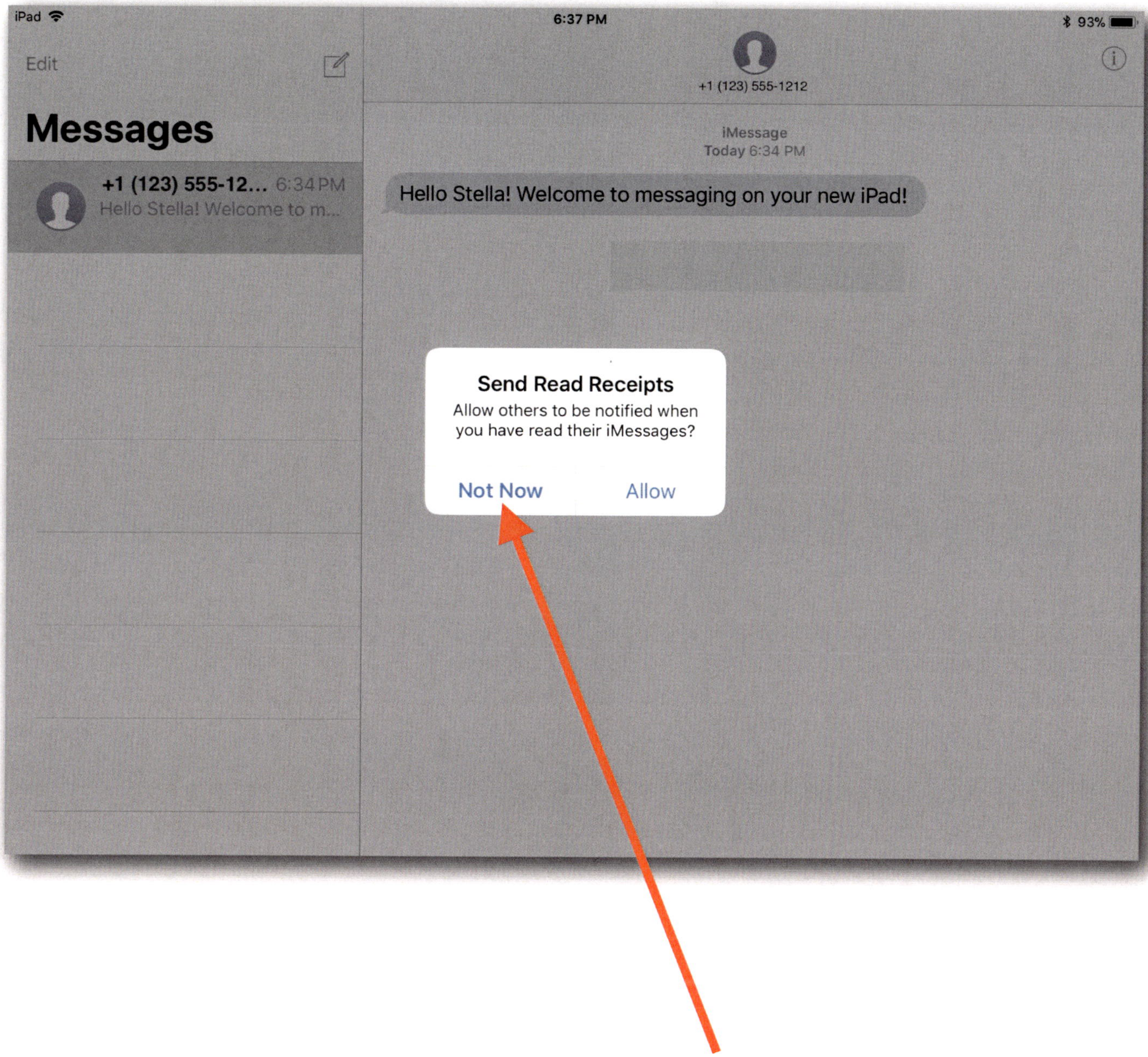
iPad
6:37 PM
93%
Edit
Messages
+1 (123) 555-12... 6:34 PM
Hello Stella! Welcome to m...
+1 (123) 555-1212
iMessage
Today 6:34 PM
Hello Stella! Welcome to messaging on your new iPad!
Send Read Receipts
Allow others to be notified when you have read their iMessages?
Not Now
Allow

If this message pops up touch here for now.

This option may be changed later.

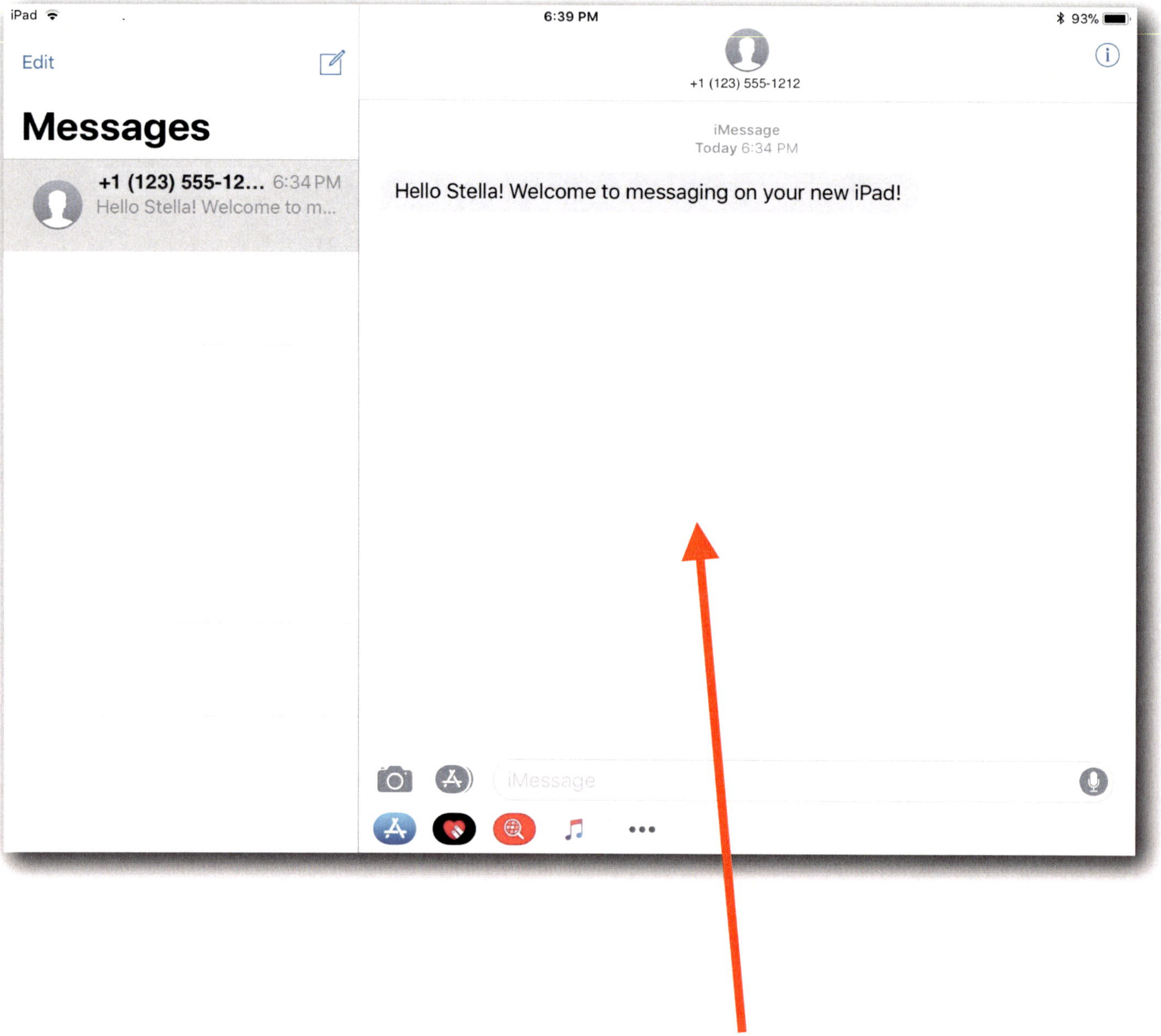
iPad
6:39 PM
93%
Edit
Messages
+1 (123) 555-12... 6:34 PM
Hello Stella! Welcome to m...
+1 (123) 555-1212
iMessage
Today 6:34 PM
Hello Stella! Welcome to messaging on your new iPad!
iMessage

The message will appear here.

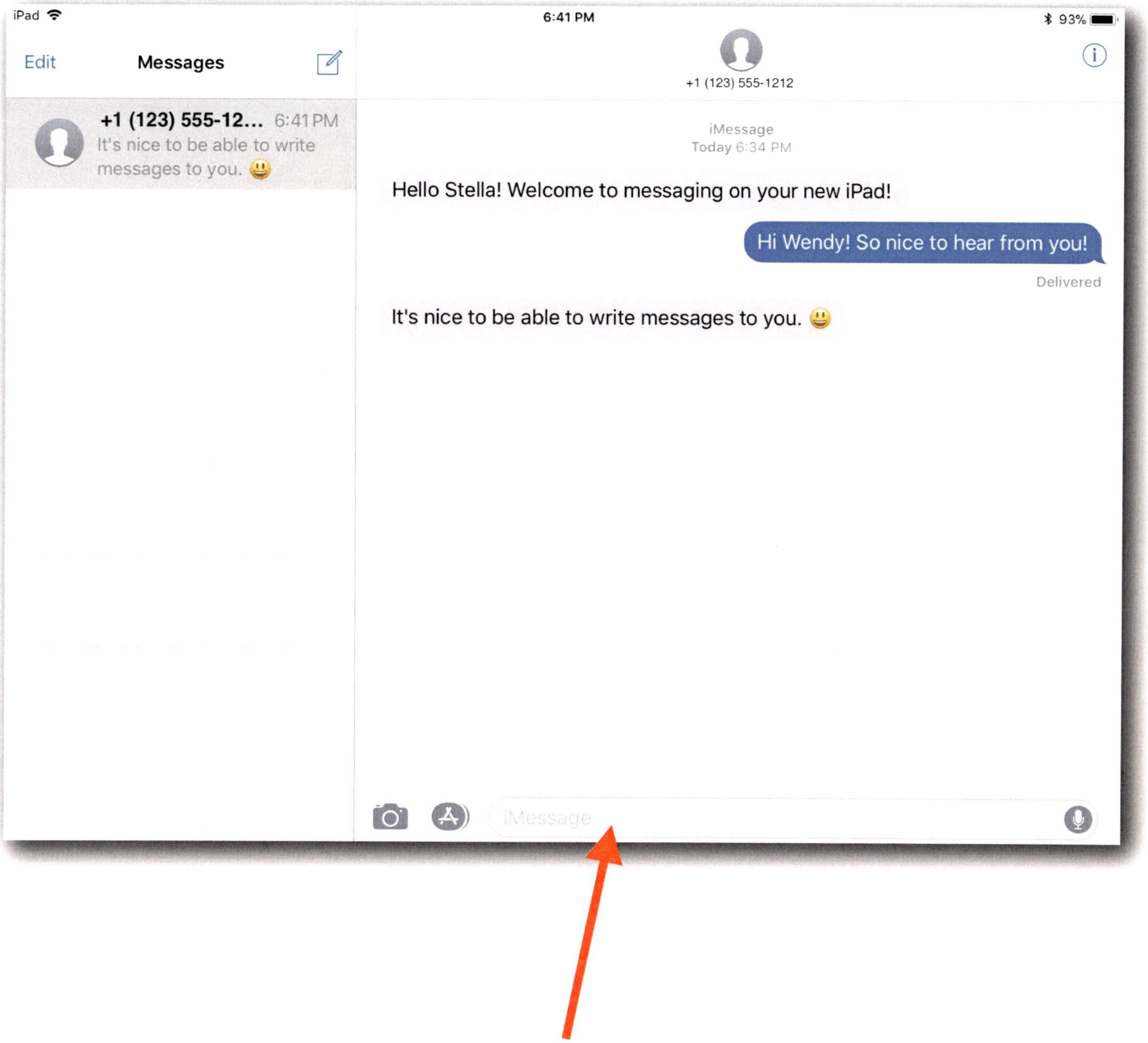
iPad
6:41 PM
93%
Edit
Messages
+1 (123) 555-12... 6:41 PM
It's nice to be able to write messages to you.
+1 (123) 555-1212
iMessage
Today 6:34 PM
Hello Stella! Welcome to messaging on your new iPad!
Hi Wendy! So nice to hear from you!
Delivered
It's nice to be able to write messages to you.
iMessage

Touch here to reply.

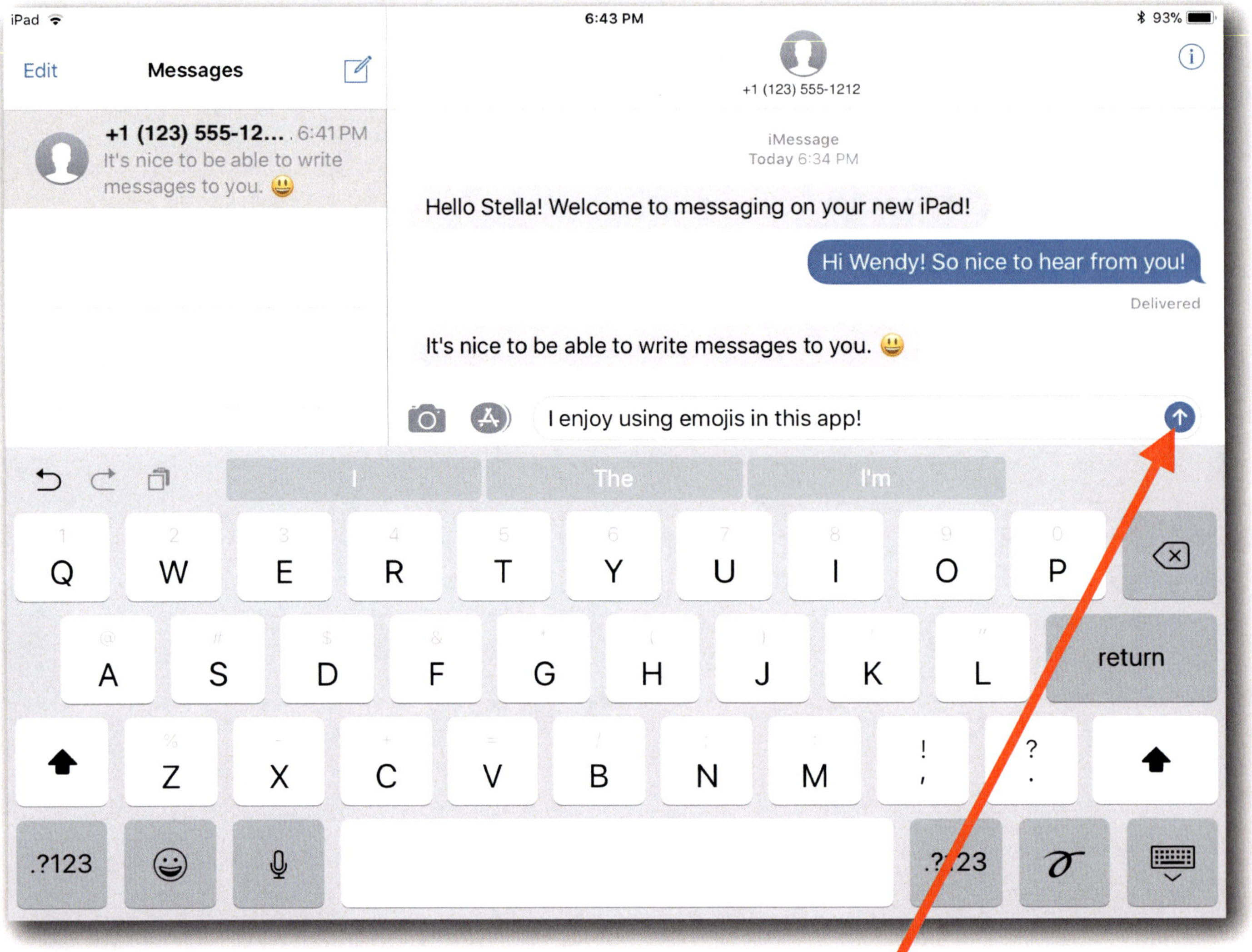
iPad
6:43 PM
93%
Edit
Messages
+1 (123) 555-12... 6:41 PM
It's nice to be able to write messages to you.
+1 (123) 555-1212
iMessage
Today 6:34 PM
Hello Stella! Welcome to messaging on your new iPad!
Hi Wendy! So nice to hear from you!
Delivered
It's nice to be able to write messages to you.
I enjoy using emojis in this app!
I
The
I'm
return
.?123

Type your message, then touch here to send it.

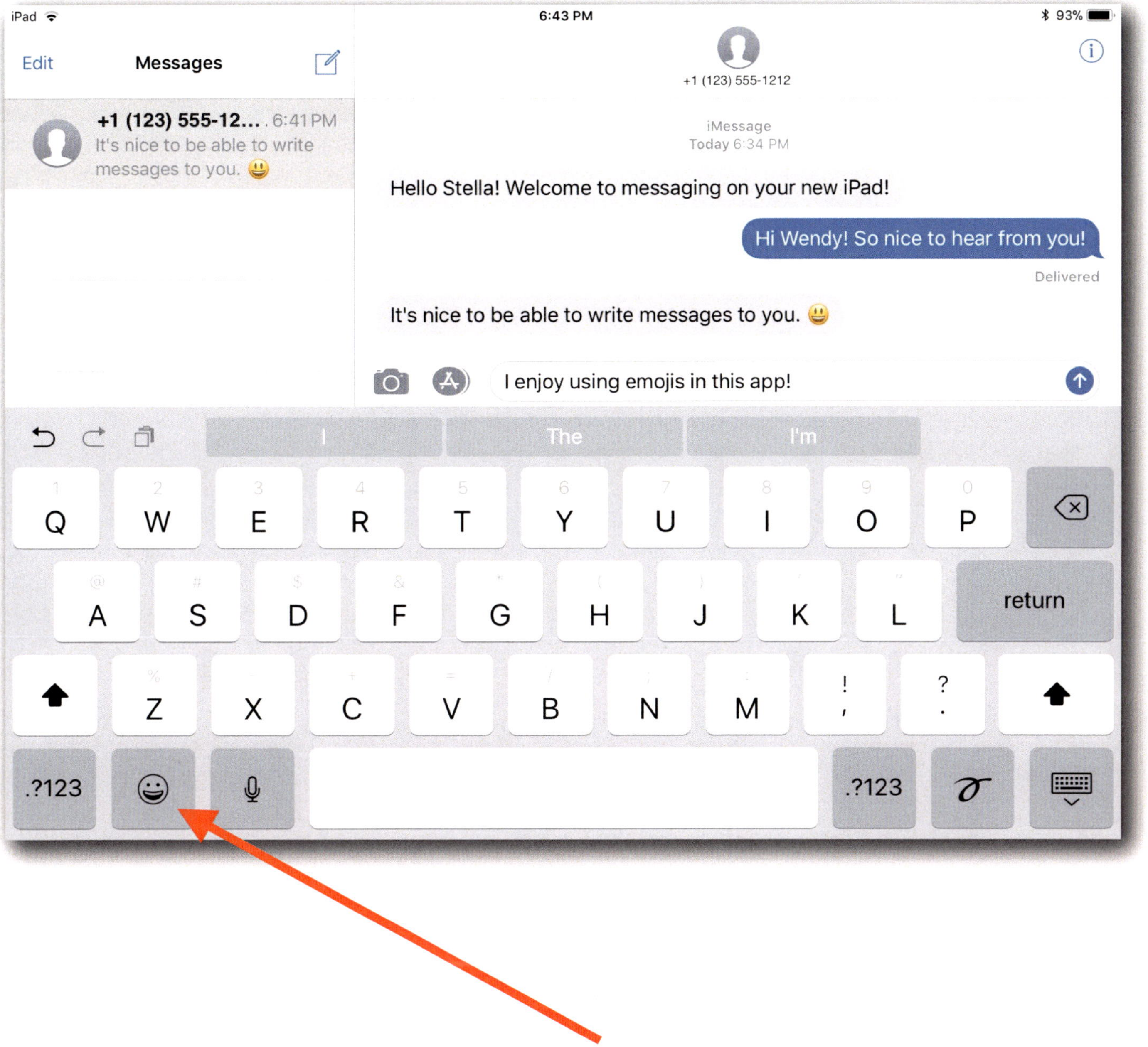
iPad
6:43 PM
93%
Edit
Messages
+1 (123) 555-1212
+1 (123) 555-12... 6:41 PM
It's nice to be able to write messages to you.
iMessage
Today 6:34 PM
Hello Stella! Welcome to messaging on your new iPad!
Hi Wendy! So nice to hear from you!
Delivered
It's nice to be able to write messages to you.
I enjoy using emojis in this app!
I
The
I'm
return
.?123
.?123

You can add fun emojis by touching here.

(Emojis are small digital images used to express emotion or ideas.)

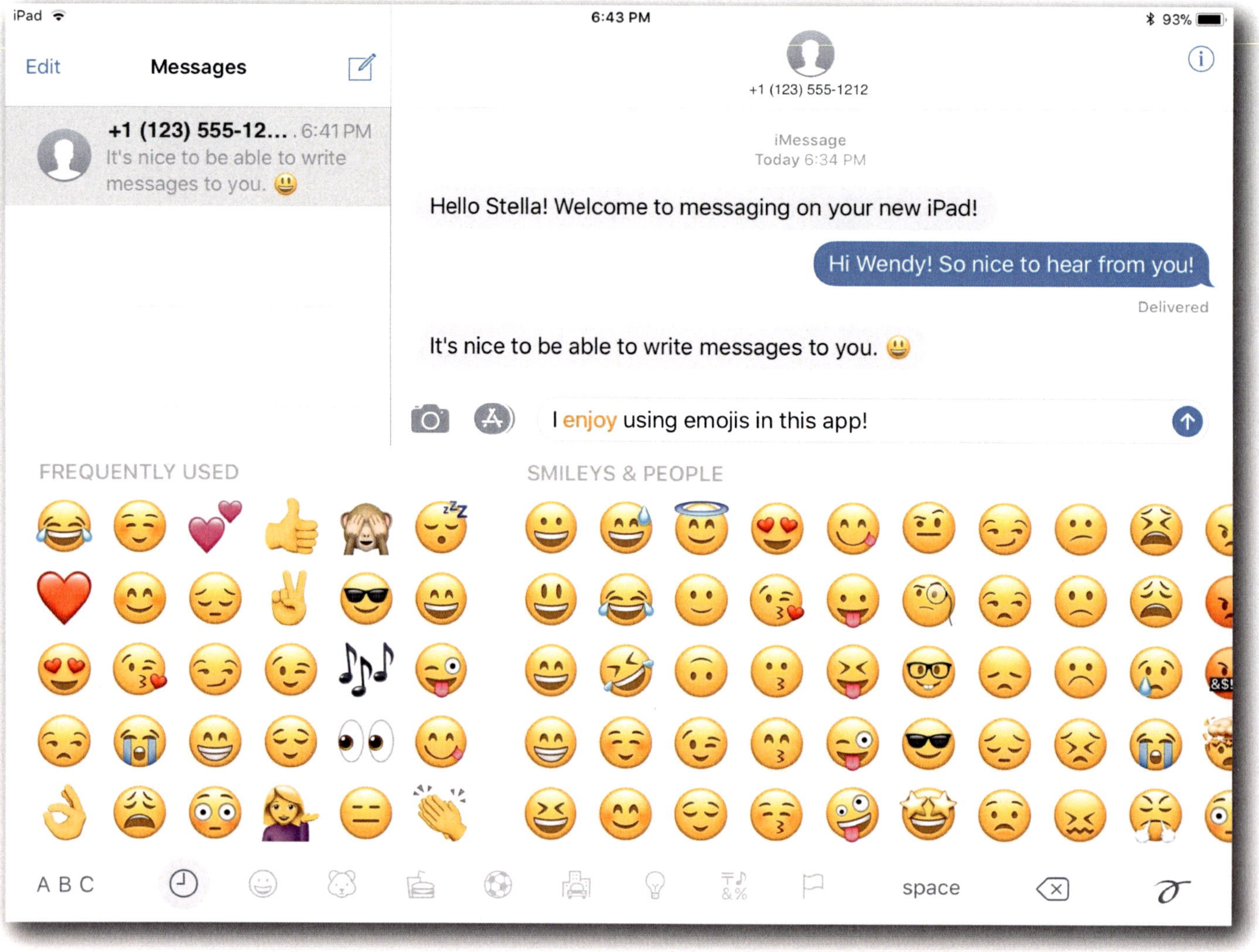
iPad
6:43 PM
93%
Edit
Messages
+1 (123) 555-12... 6:41 PM
It's nice to be able to write messages to you.
+1 (123) 555-1212
iMessage
Today 6:34 PM
Hello Stella! Welcome to messaging on your new iPad!
Hi Wendy! So nice to hear from you!
Delivered
It's nice to be able to write messages to you.
I enjoy using emojis in this app!
FREQUENTLY USED
SMILEYS & PEOPLE
A B C
space

Touch the emoji you want to use and it will appear in your message field.

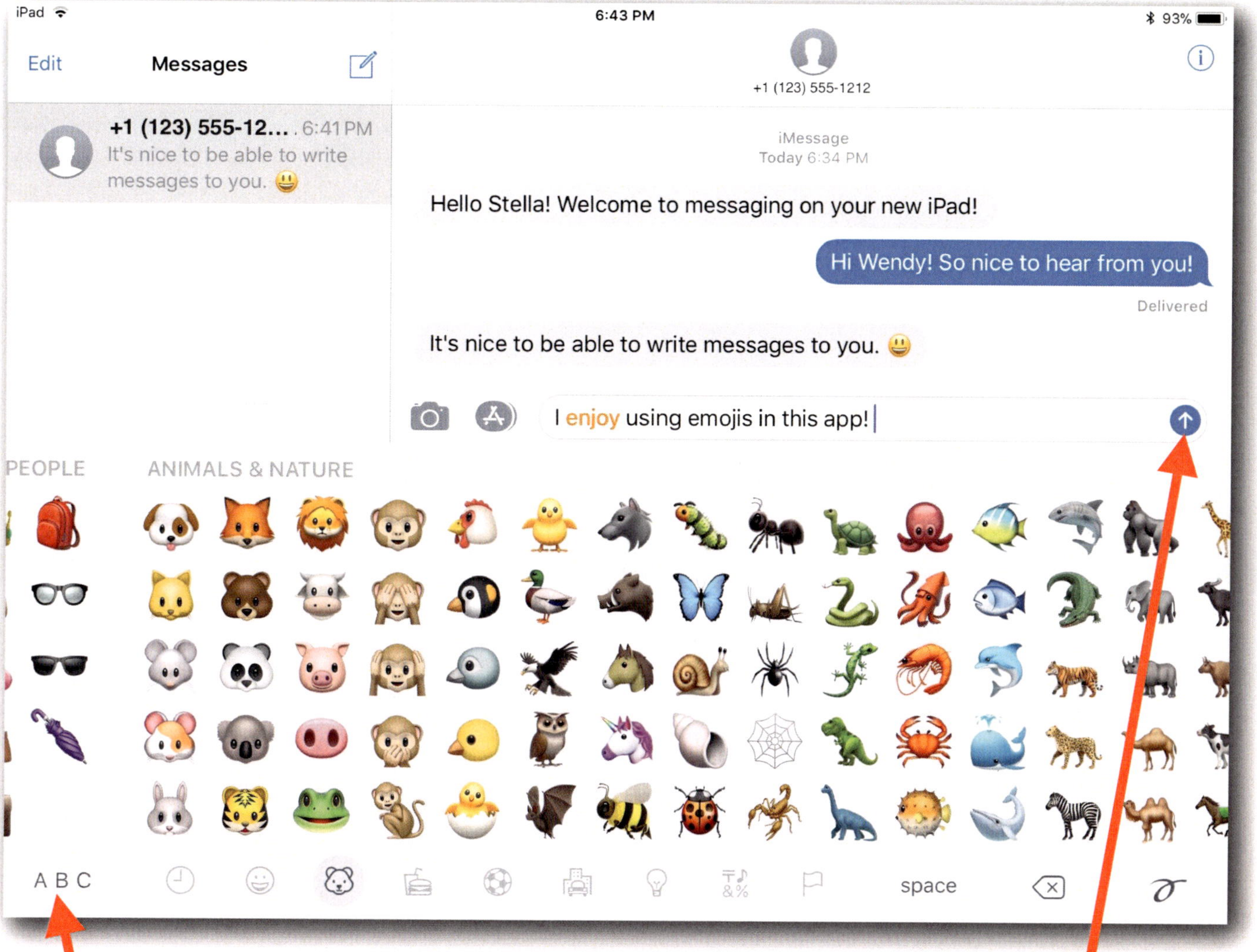
iPad
6:43 PM
93%
Edit
Messages
+1 (123) 555-1212
+1 (123) 555-12... 6:41 PM
It's nice to be able to write messages to you.
iMessage
Today 6:34 PM
Hello Stella! Welcome to messaging on your new iPad!
Hi Wendy! So nice to hear from you!
Delivered
It's nice to be able to write messages to you.
I enjoy using emojis in this app!
PEOPLE
ANIMALS & NATURE
A B C
space

You can scroll left or right through many emojis.

Touch here to return to the regular keyboard.

Touch here
when you're ready to send your message.

Writing A New Text Message
3 Methods

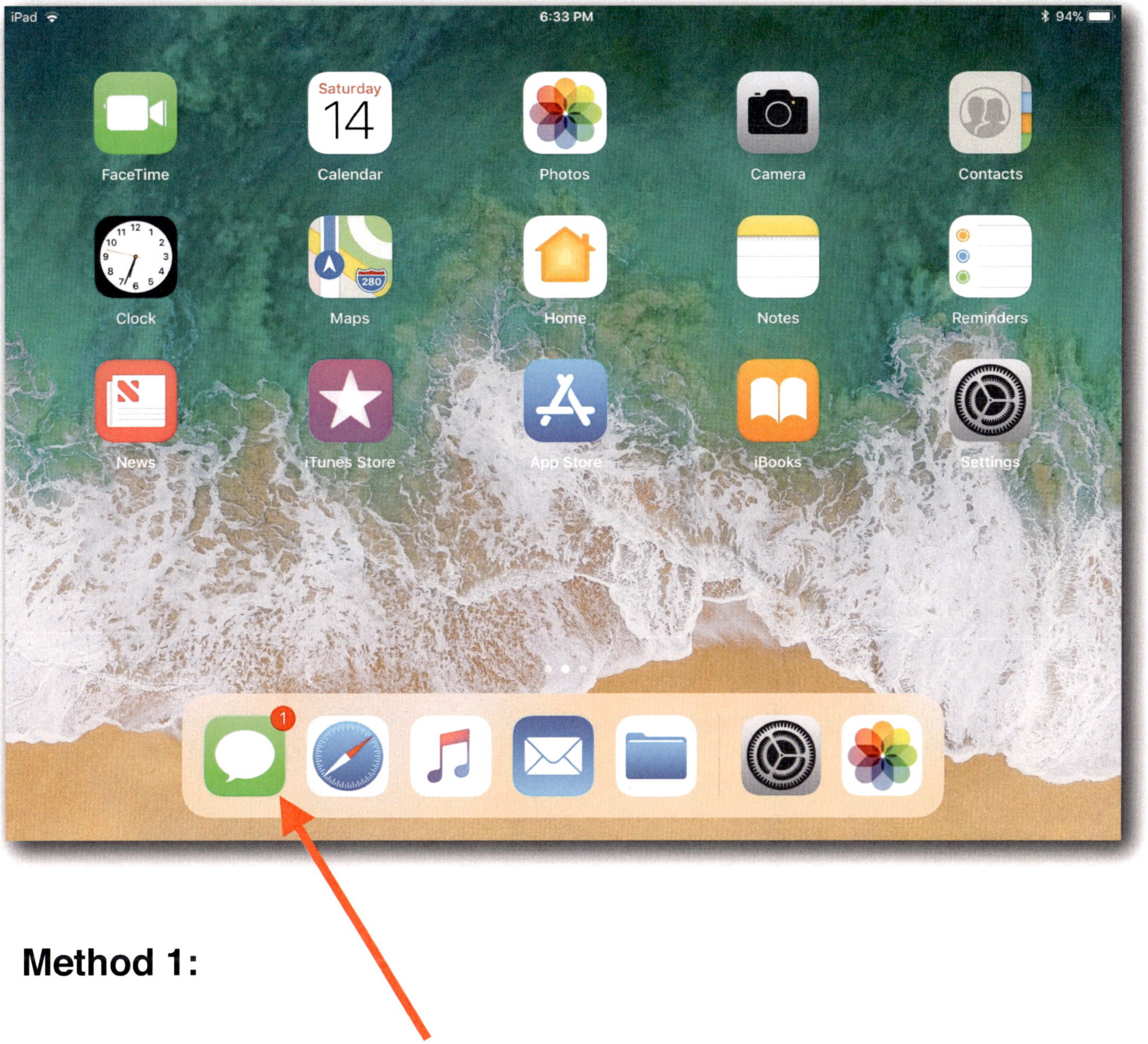

Method 1:

Touch here to open Text Messaging.

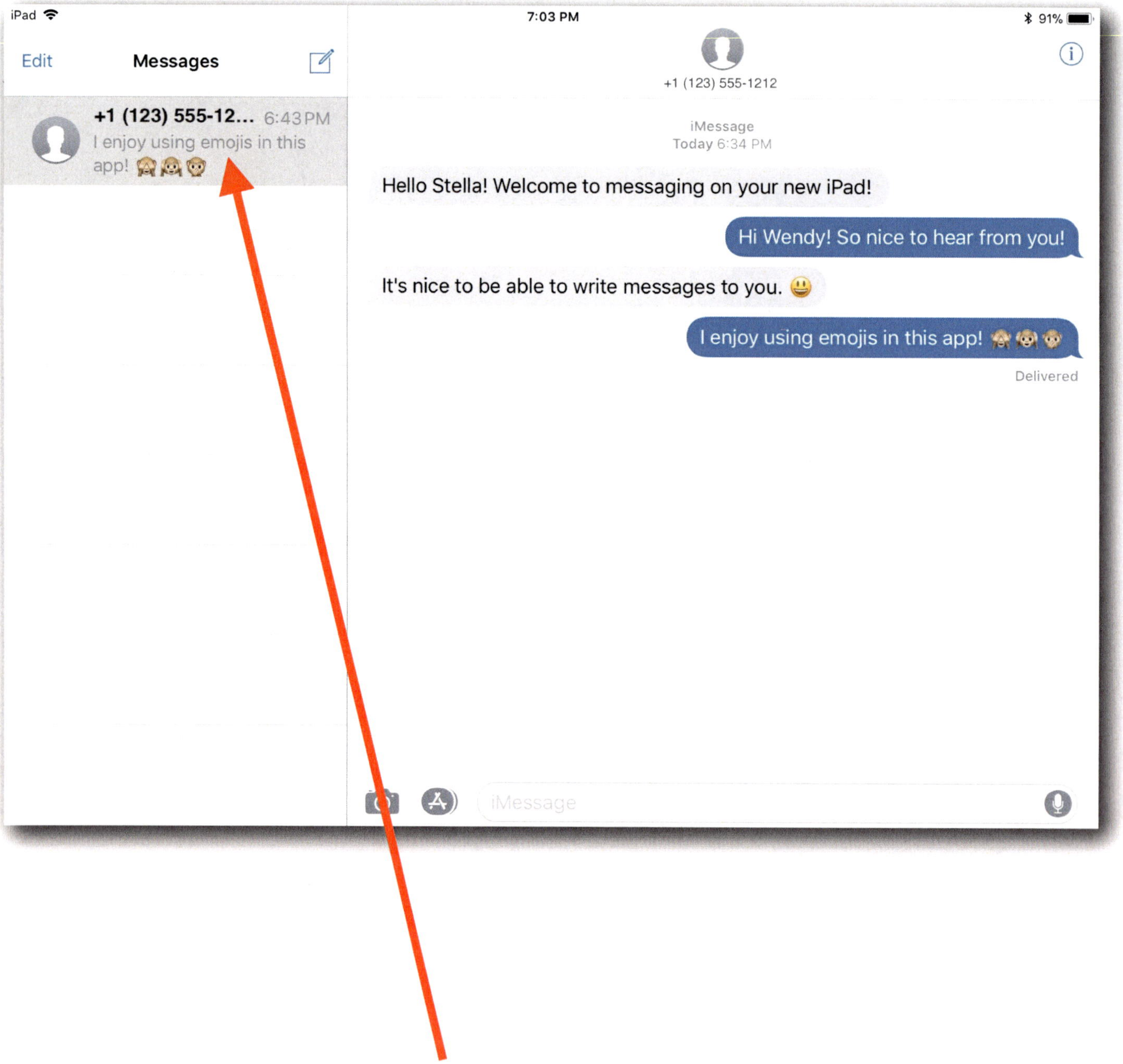
iPad
7:03 PM
91%
Edit
Messages
+1 (123) 555-12... 6:43 PM
I enjoy using emojis in this app!
+1 (123) 555-1212
iMessage
Today 6:34 PM
Hello Stella! Welcome to messaging on your new iPad!
Hi Wendy! So nice to hear from you!
It's nice to be able to write messages to you.
I enjoy using emojis in this app!
Delivered
iMessage

Touch the conversation you'd like to continue.

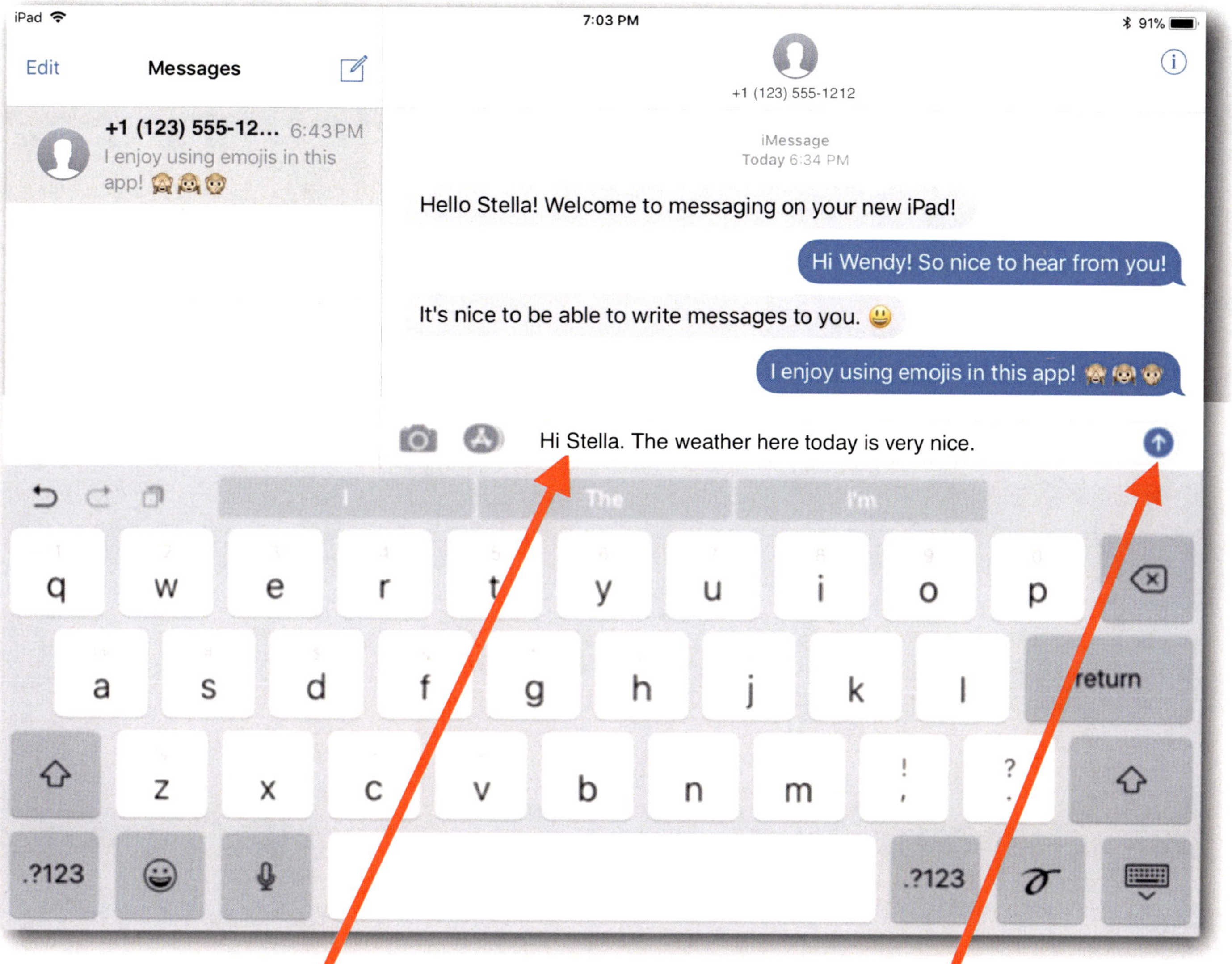
iPad
7:03 PM
91%
Edit
Messages
+1 (123) 555-1212
+1 (123) 555-12... 6:43 PM
I enjoy using emojis in this app!
iMessage
Today 6:34 PM
Hello Stella! Welcome to messaging on your new iPad!
Hi Wendy! So nice to hear from you!
It's nice to be able to write messages to you.
I enjoy using emojis in this app!
Hi Stella. The weather here today is very nice.
return
.?123

Touch here and then type your text message.

Touch here to send.

Press the Home Button when you're done.

Method 2:

Touch here to open Text Messaging.

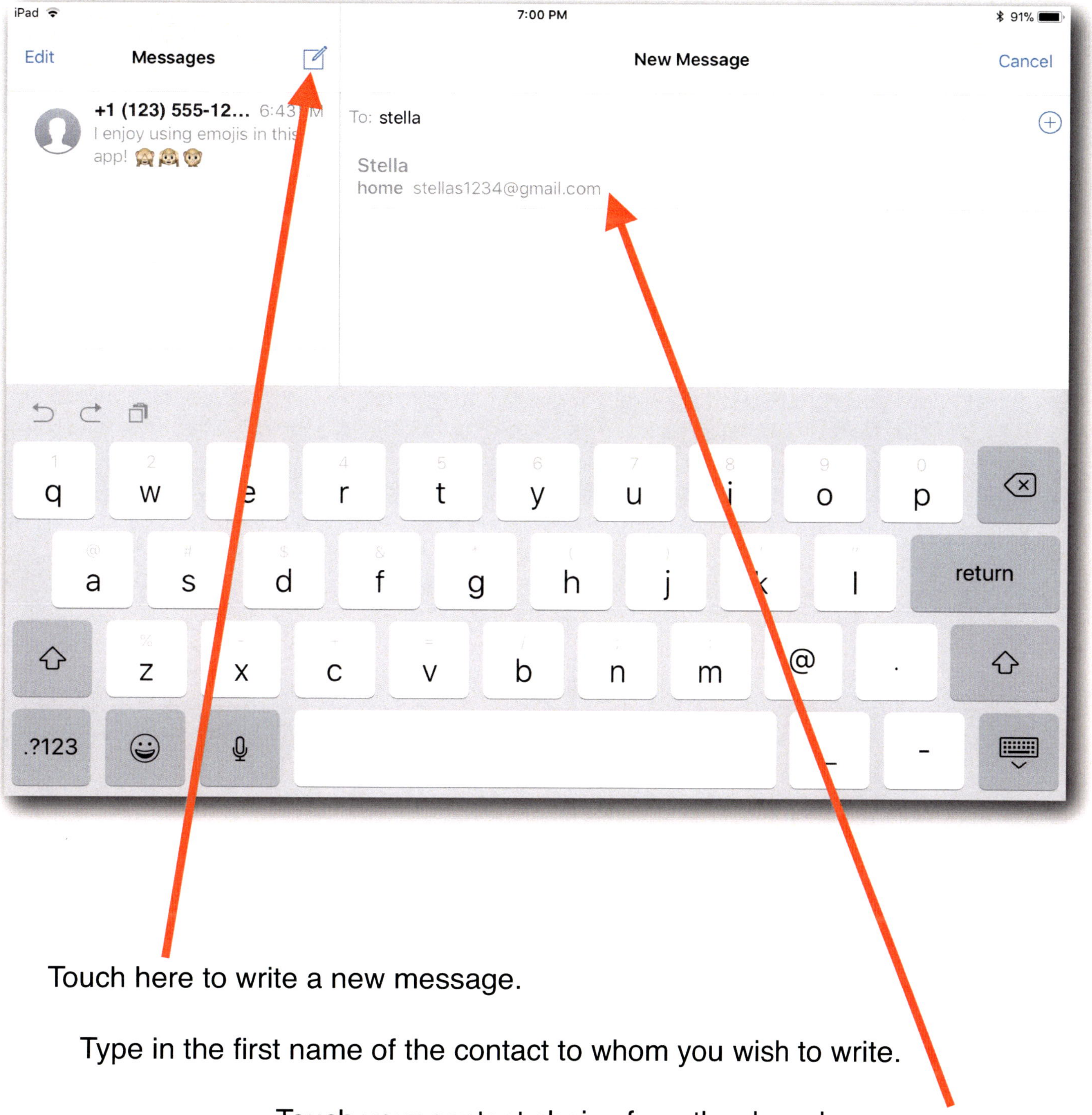

Touch here to write a new message.

Type in the first name of the contact to whom you wish to write.

Touch your contact choice from the drop-down menu.

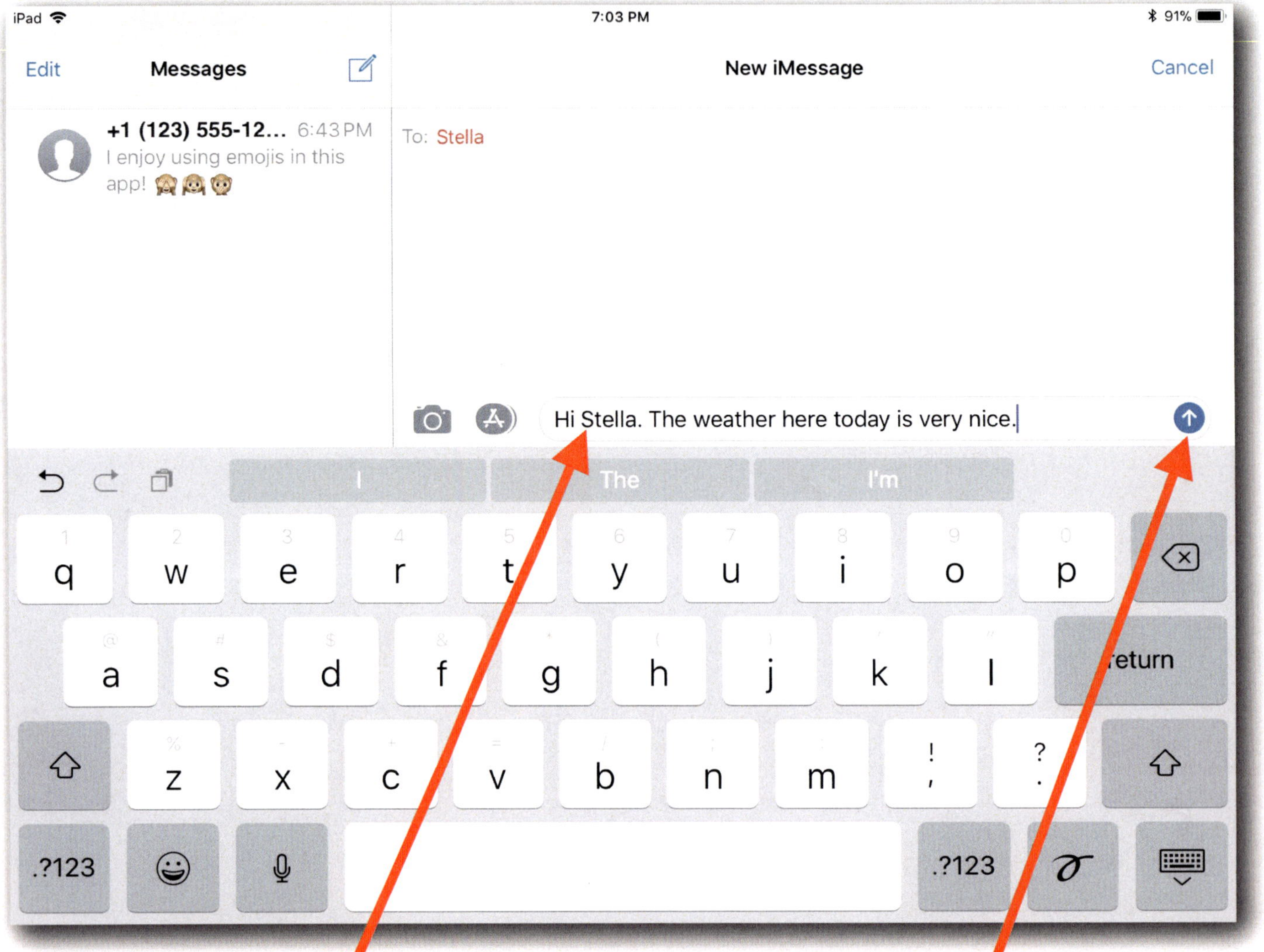

Touch here and then type your text message.

Touch here to send.

Press the Home Button when you're done.

Method 3

Touch here to open Contacts.

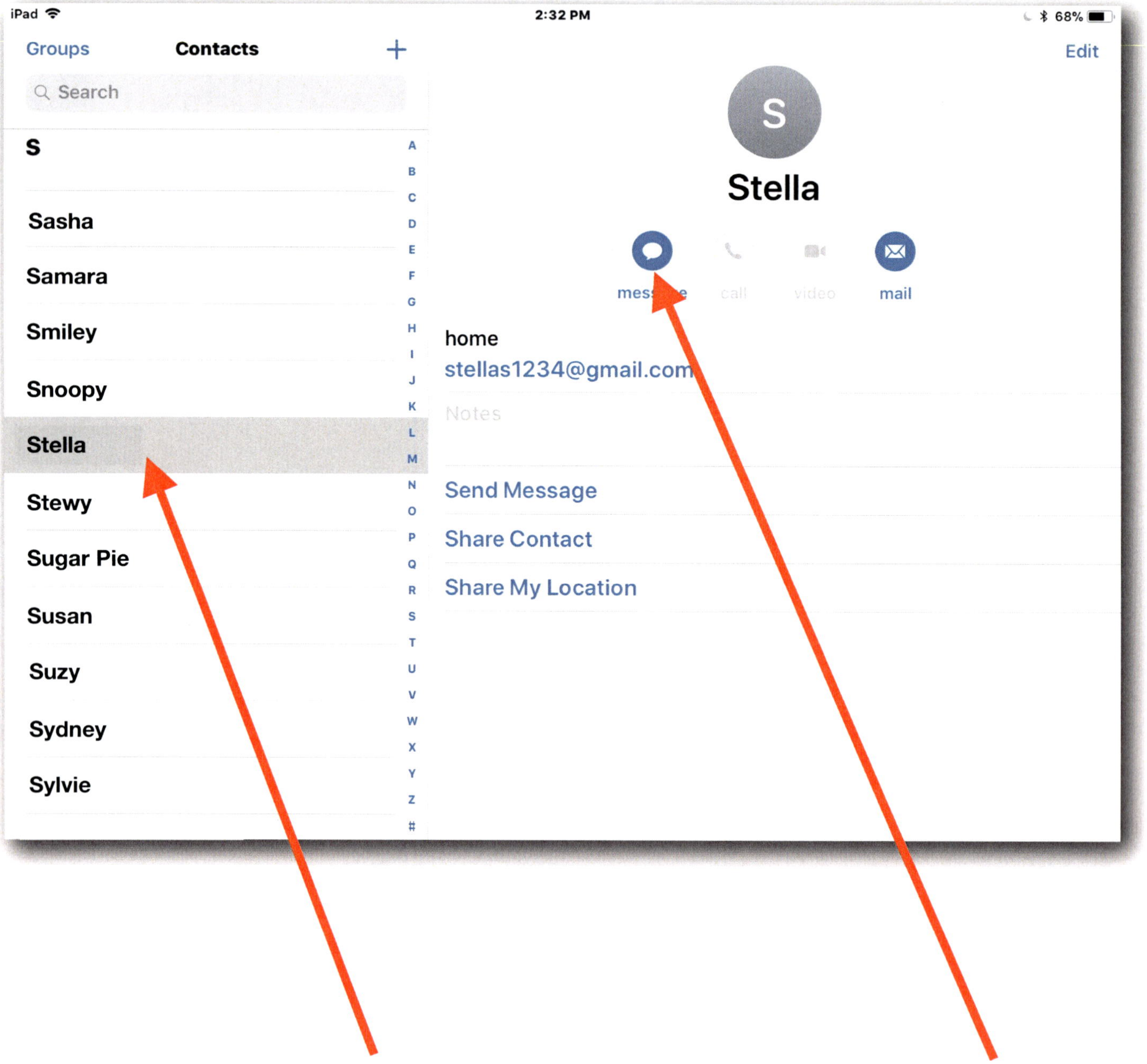

Touch your contact choice.

Then touch here.

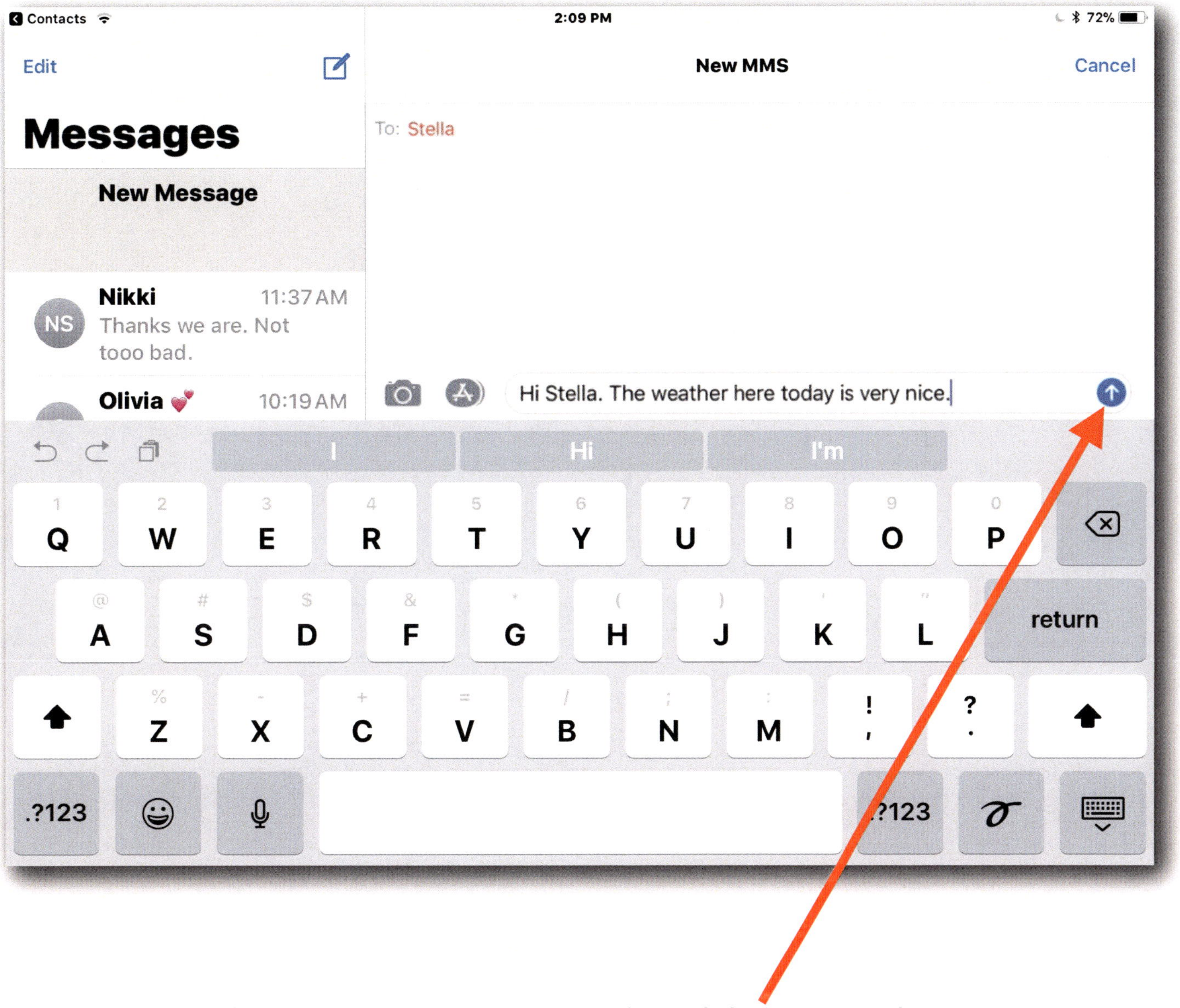

Type your text message and touch here to send.

Press the Home Button when you're done.

Using Email

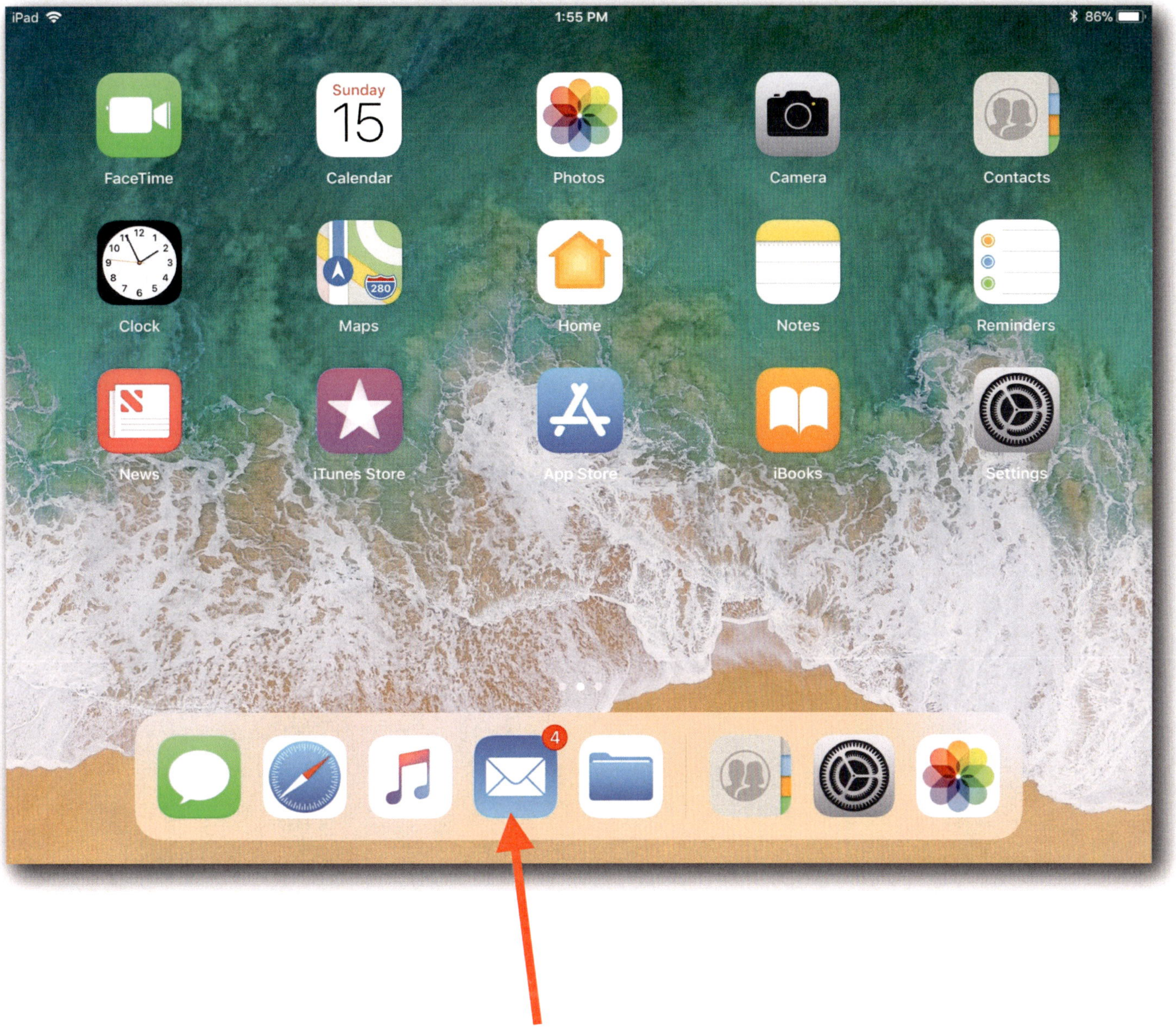

Touch here to open email.

Reading And Replying To Emails

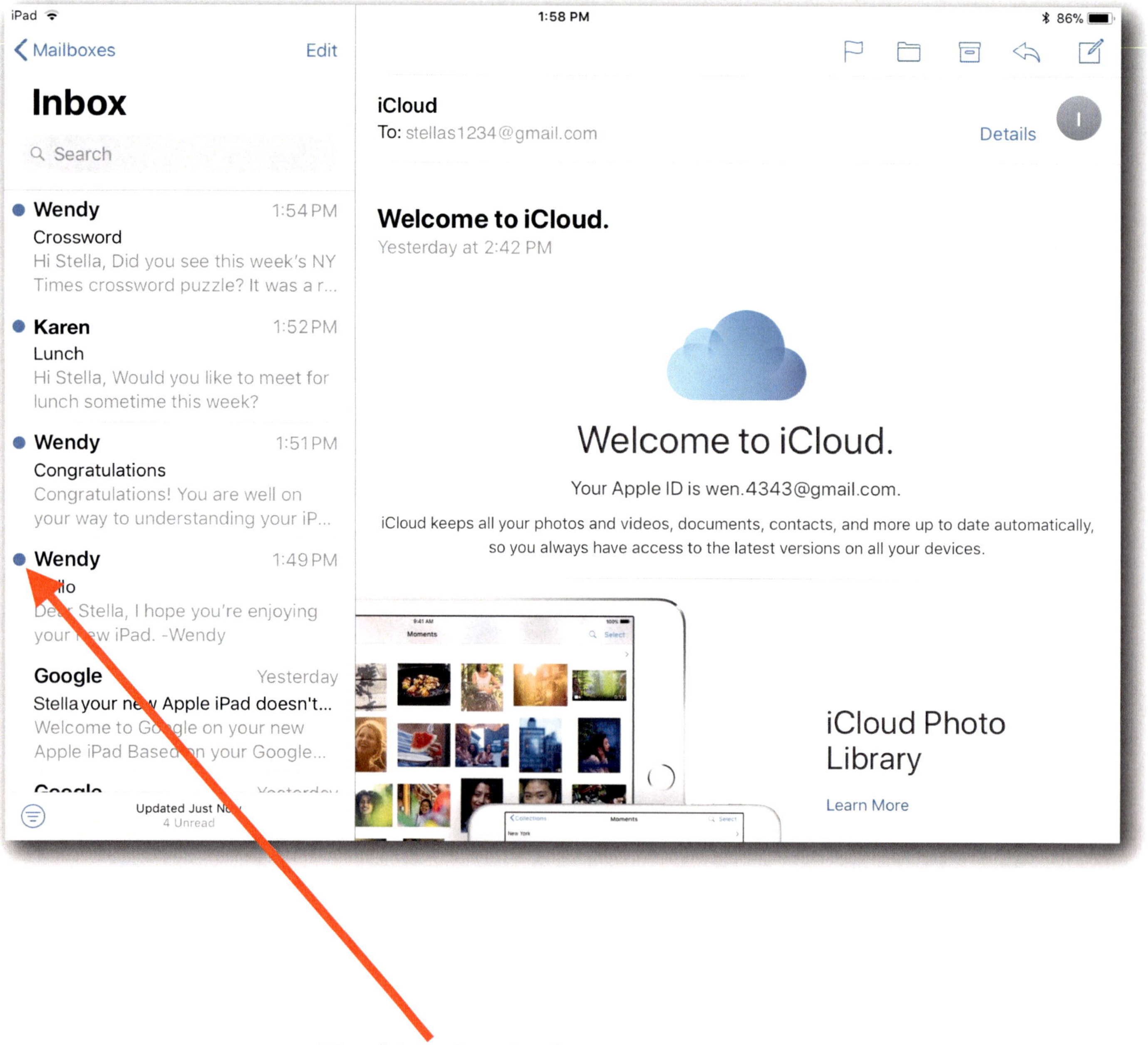

iPad
1:58 PM
86%
Mailboxes
Edit
Inbox
Search
Wendy 1:54 PM
Crossword
Hi Stella, Did you see this week's NY Times crossword puzzle? It was a r...
Karen 1:52 PM
Lunch
Hi Stella, Would you like to meet for lunch sometime this week?
Wendy 1:51 PM
Congratulations
Congratulations! You are well on your way to understanding your iP...
Wendy 1:49 PM
Google Yesterday
Stella your new Apple iPad doesn't...
Updated Just Now
4 Unread
iCloud
To: stellas1234@gmail.com
Details
Welcome to iCloud.
Yesterday at 2:42 PM
Welcome to iCloud.
Your Apple ID is wen.4343@gmail.com.
iCloud keeps all your photos and videos, documents, contacts, and more up to date automatically, so you always have access to the latest versions on all your devices.
iCloud Photo Library
Learn More

The blue dots indicate unread emails.

Touch an email you'd like to open and read.

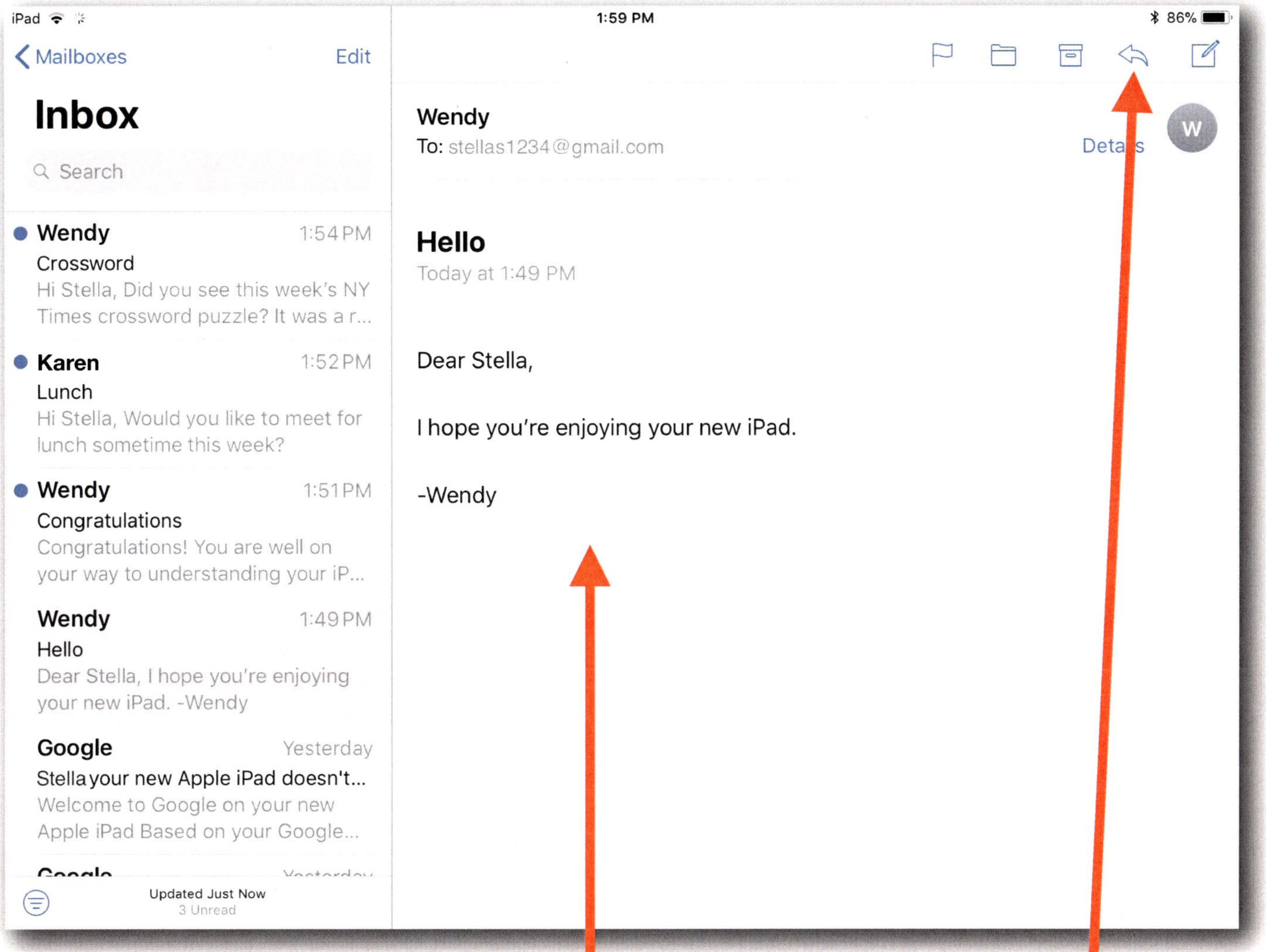
iPad
1:59 PM
86%
Mailboxes
Edit
Inbox
Search
Wendy
1:54 PM
Crossword
Hi Stella, Did you see this week's NY Times crossword puzzle? It was a r...
Karen
1:52 PM
Lunch
Hi Stella, Would you like to meet for lunch sometime this week?
Wendy
1:51 PM
Congratulations
Congratulations! You are well on your way to understanding your iP...
Wendy
1:49 PM
Hello
Dear Stella, I hope you're enjoying your new iPad. -Wendy
Google
Yesterday
Stella your new Apple iPad doesn't...
Welcome to Google on your new Apple iPad Based on your Google...
Updated Just Now
3 Unread
Wendy
To: stellas1234@gmail.com
Details
W
Hello
Today at 1:49 PM
Dear Stella,
I hope you're enjoying your new iPad.
-Wendy

The chosen email will appear on the right side of the screen.

Touch here
to reply to the email.

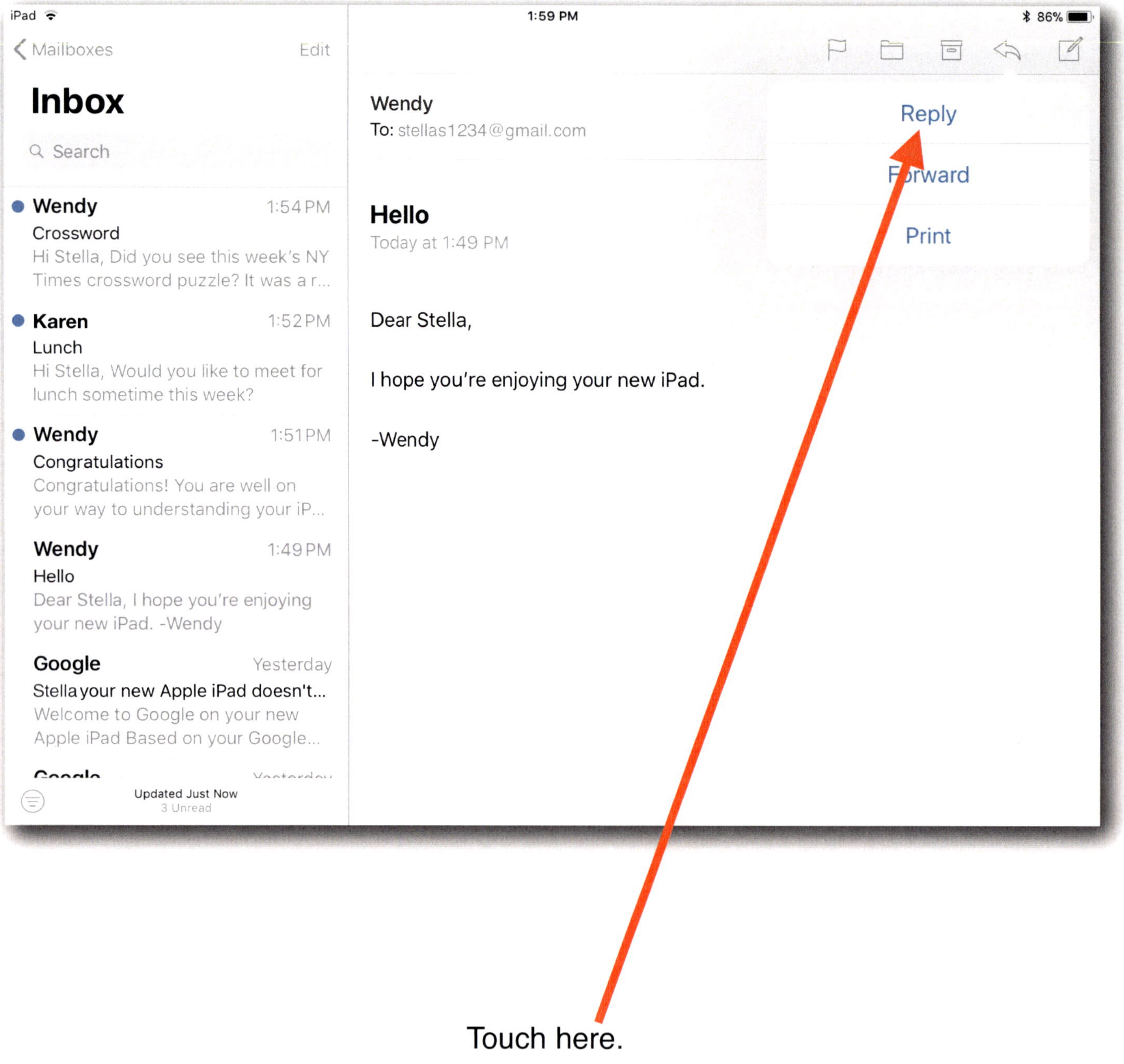
iPad
1:59 PM
86%
Mailboxes
Edit
Inbox
Search
Wendy
1:54 PM
Crossword
Hi Stella, Did you see this week's NY Times crossword puzzle? It was a r...
Karen
1:52 PM
Lunch
Hi Stella, Would you like to meet for lunch sometime this week?
Wendy
1:51 PM
Congratulations
Congratulations! You are well on your way to understanding your iP...
Wendy
1:49 PM
Hello
Dear Stella, I hope you're enjoying your new iPad. -Wendy
Google
Yesterday
Stella your new Apple iPad doesn't...
Welcome to Google on your new Apple iPad Based on your Google...
Updated Just Now
3 Unread
Wendy
To: stellas1234@gmail.com
Hello
Today at 1:49 PM
Dear Stella,
I hope you're enjoying your new iPad.
-Wendy
Reply
Forward
Print
Touch here.

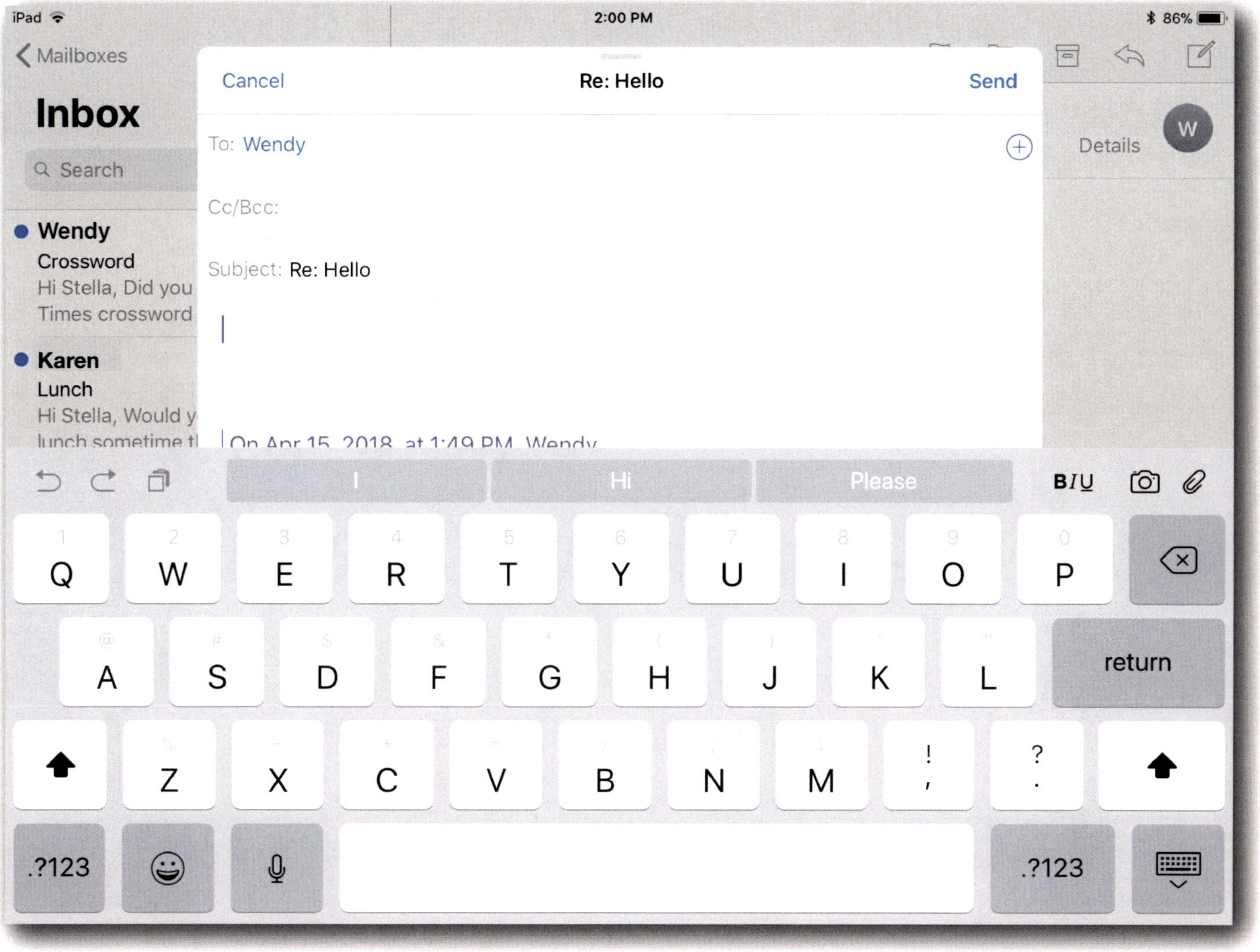

Type your email message.

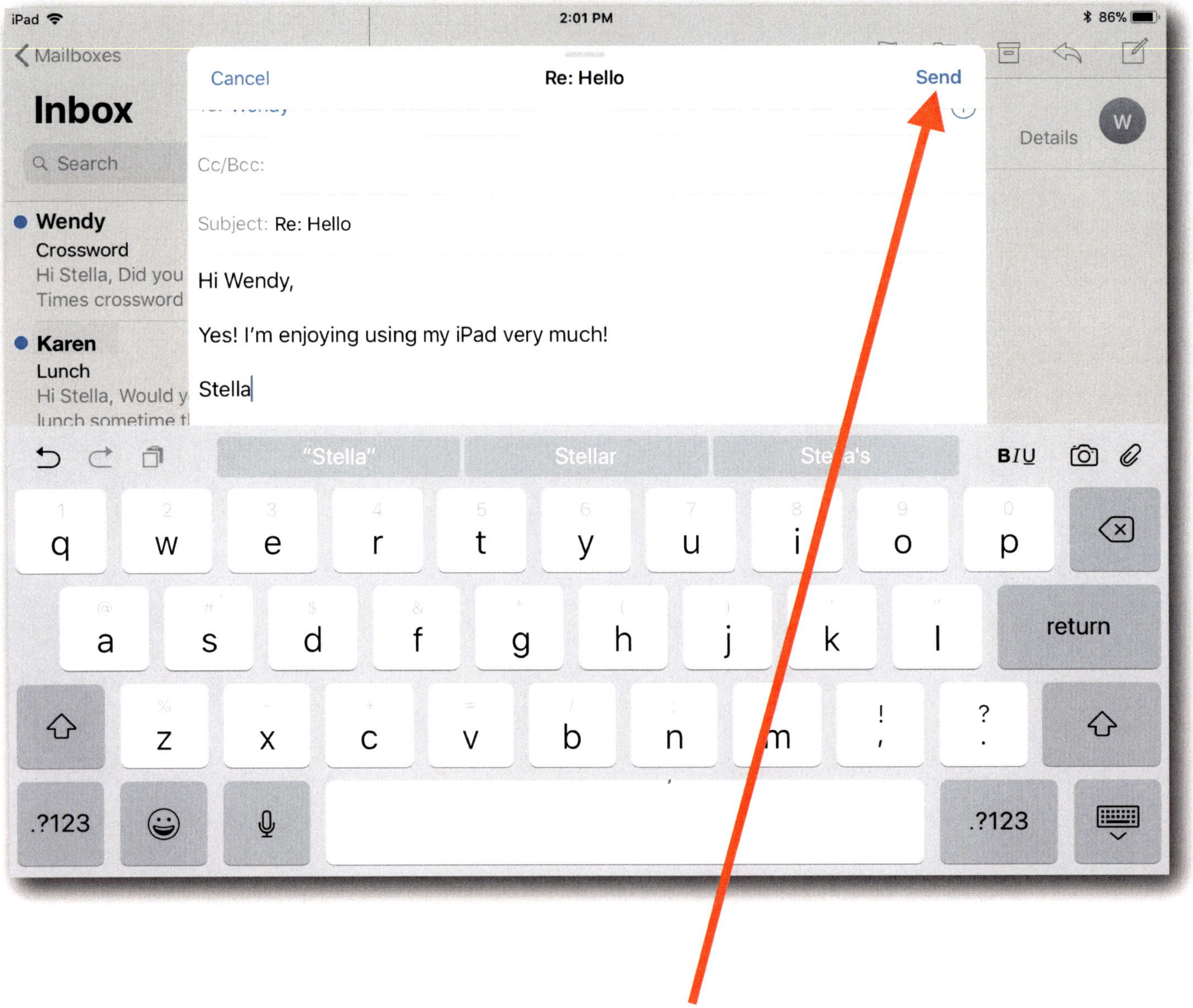

When you're done typing, touch here to send your email.

Press the Home Button when you're done.

Writing A New Email

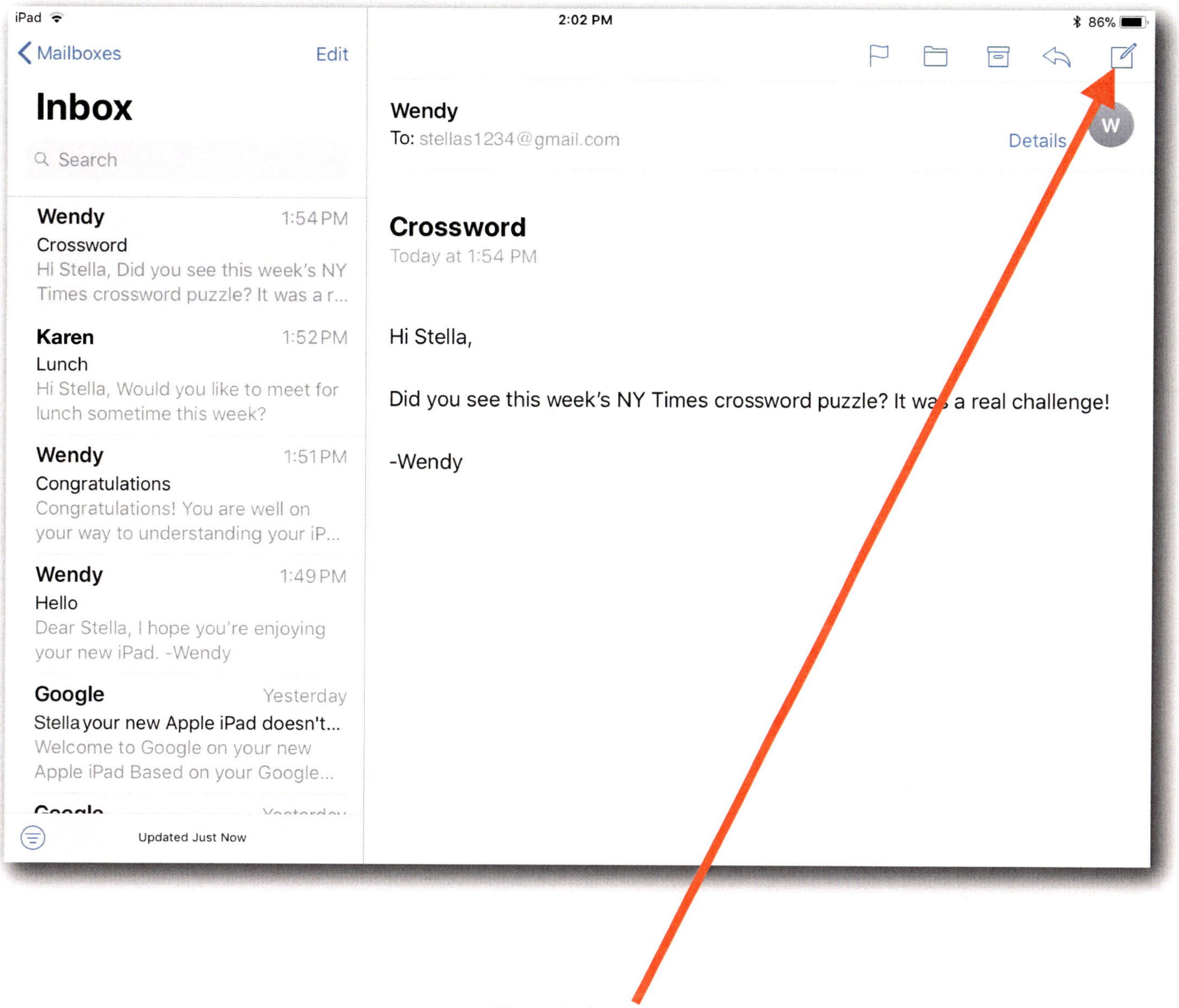
iPad
2:02 PM
86%
Mailboxes
Edit
Inbox
Search
Wendy
1:54 PM
Crossword
Hi Stella, Did you see this week's NY Times crossword puzzle? It was a r...
Karen
1:52 PM
Lunch
Hi Stella, Would you like to meet for lunch sometime this week?
Wendy
1:51 PM
Congratulations
Congratulations! You are well on your way to understanding your iP...
Wendy
1:49 PM
Hello
Dear Stella, I hope you're enjoying your new iPad. -Wendy
Google
Yesterday
Stella your new Apple iPad doesn't...
Welcome to Google on your new Apple iPad Based on your Google...
Updated Just Now
Wendy
To: stellas1234@gmail.com
Details
W
Crossword
Today at 1:54 PM
Hi Stella,
Did you see this week's NY Times crossword puzzle? It was a real challenge!
-Wendy

Touch here.

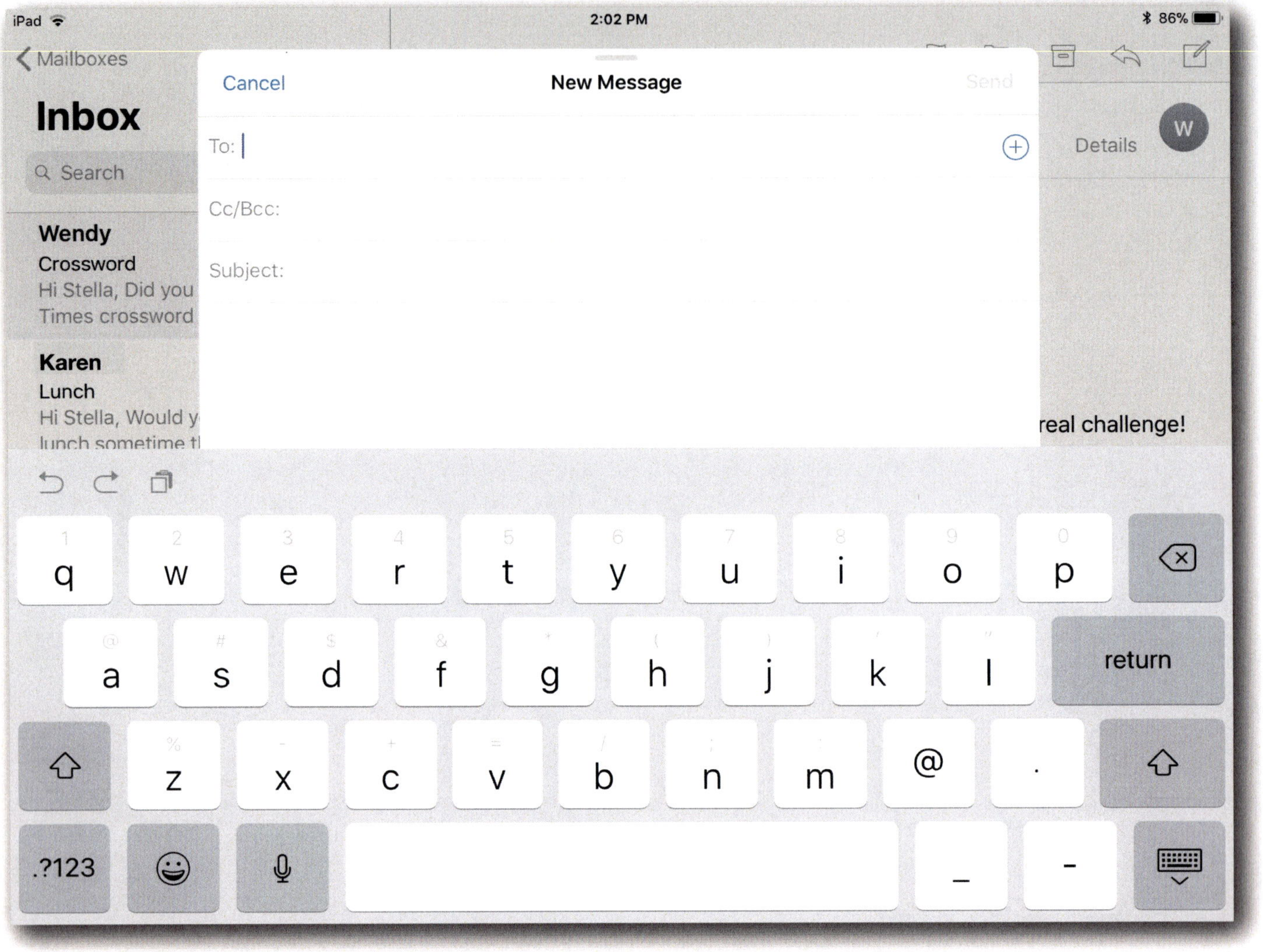

Type the first few letters of the person you wish to email.

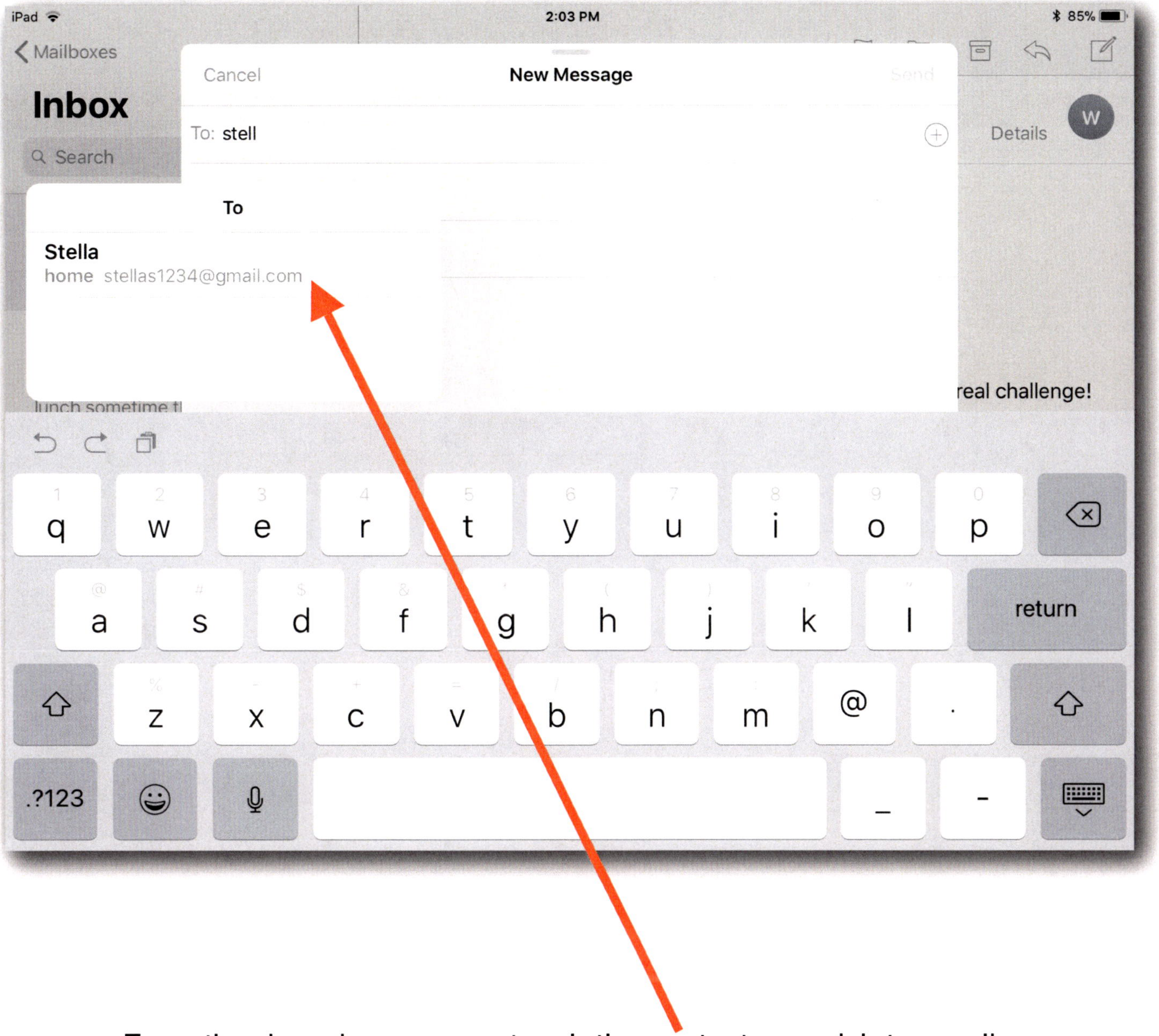

From the drop down menu, touch the contact you wish to email.

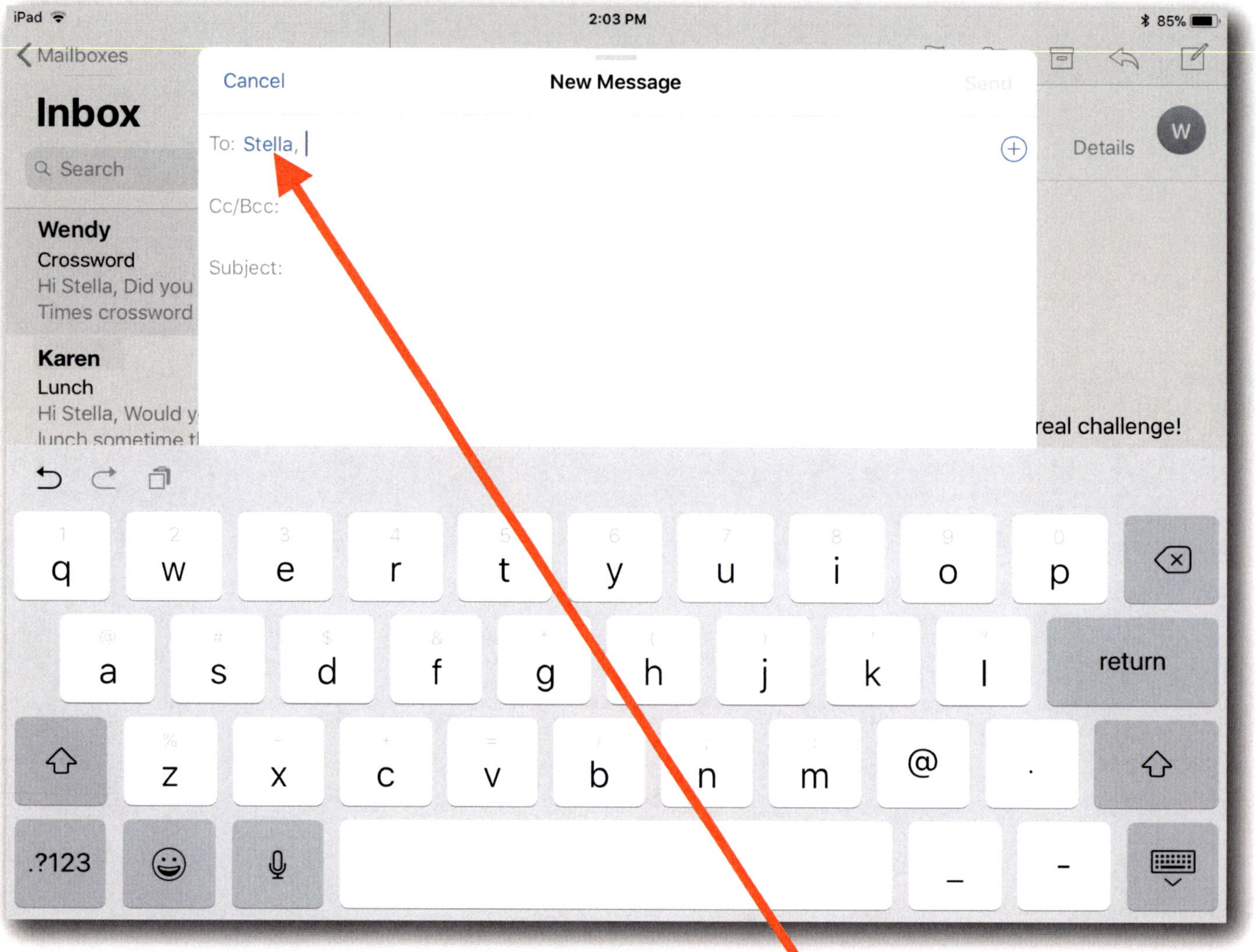

Your chosen contact's name will show up here.

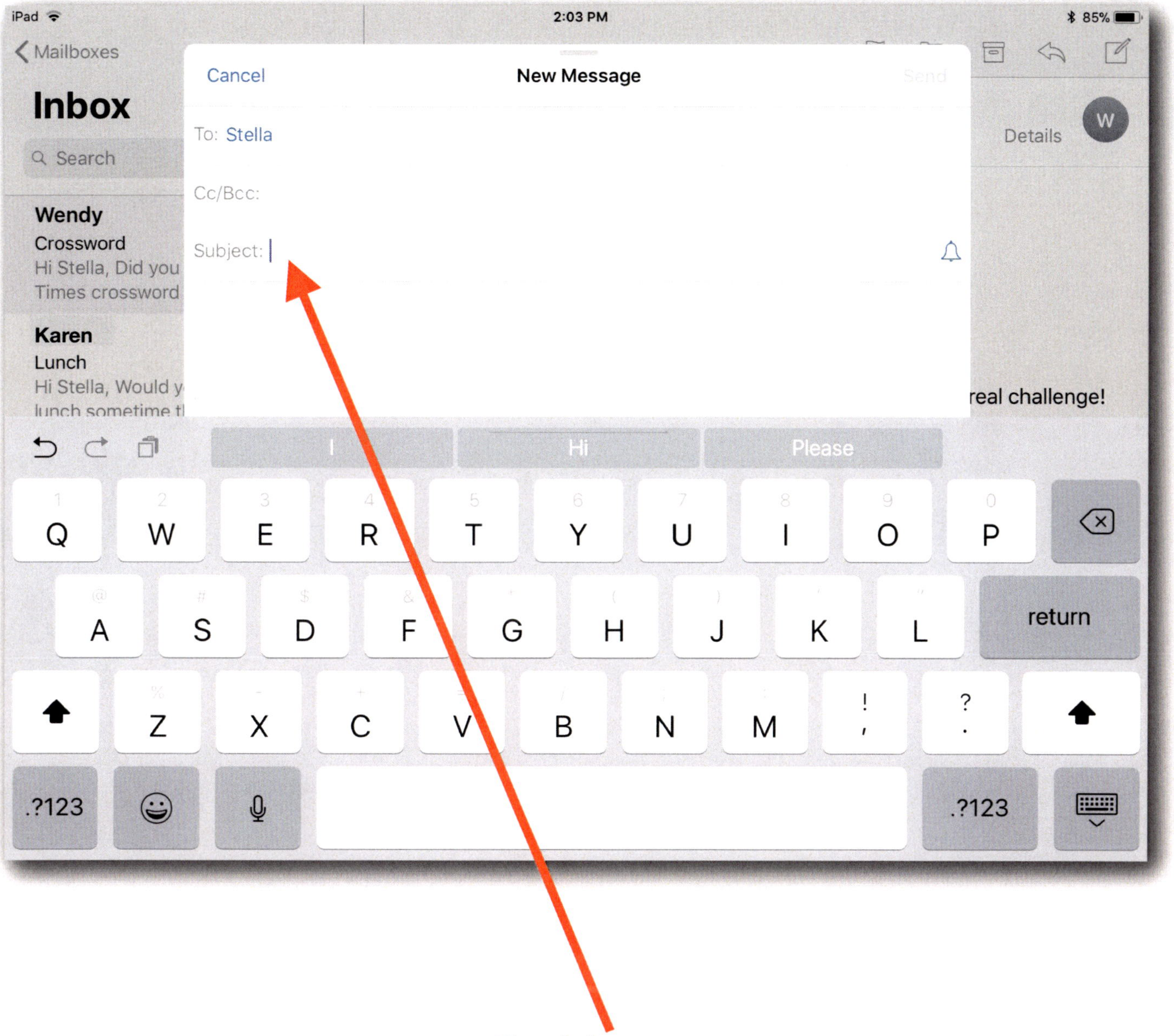

Touch here.

The blue line (cursor) will blink to indicate where your typing will appear.

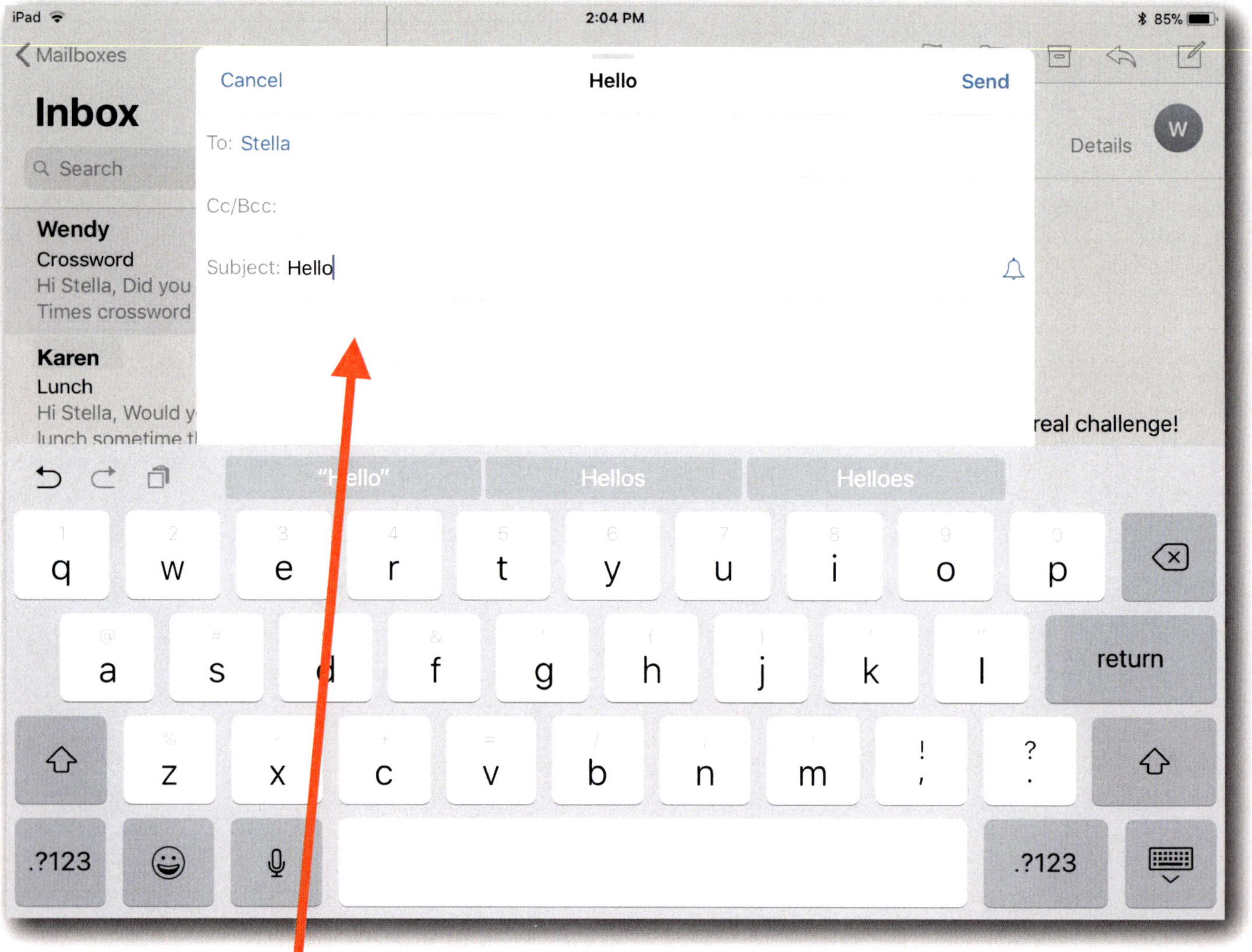

Type in a subject for your email.

Note: Emails typically require a subject.

Next, touch here,
in the body-of-the-email field.

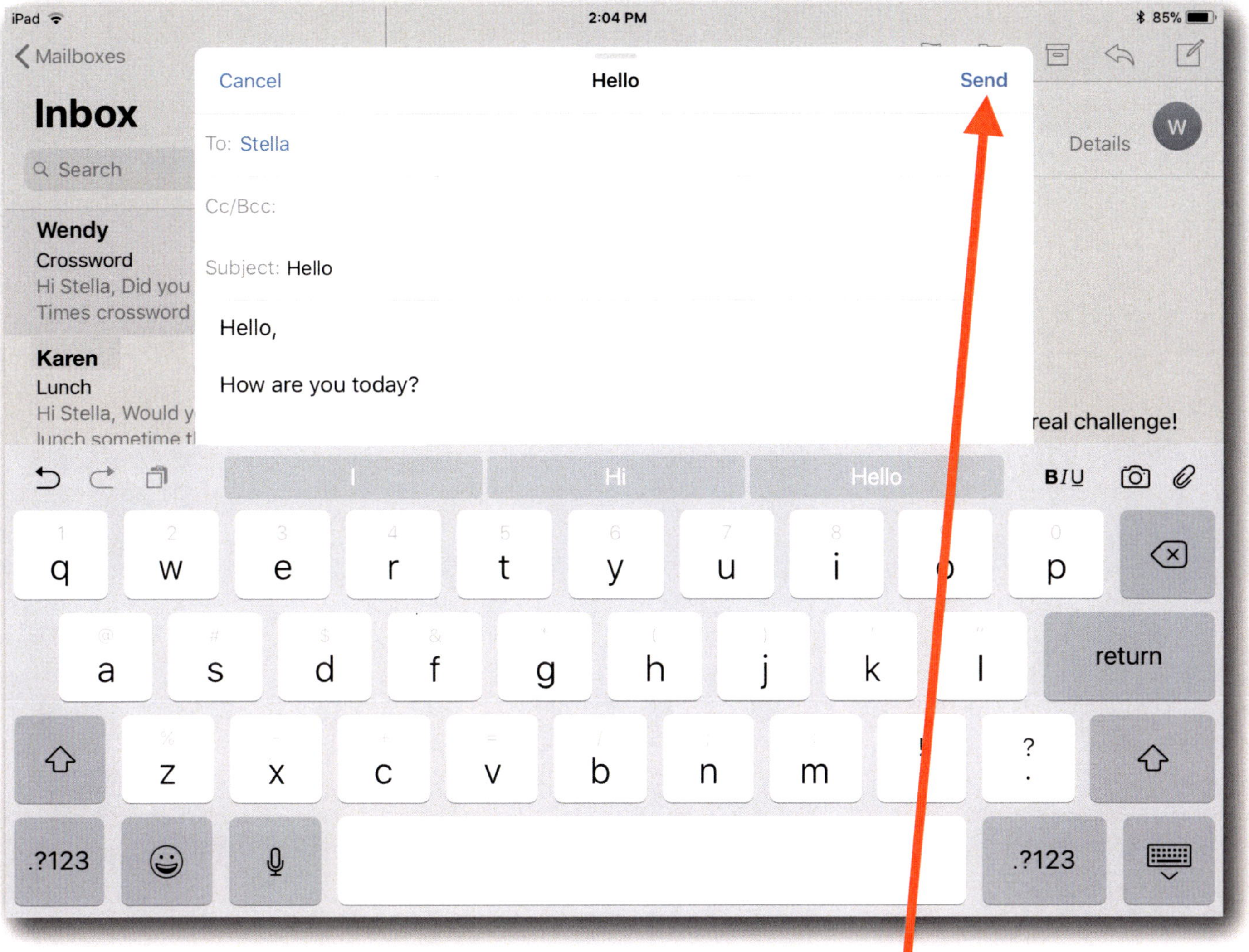

Type your email message.

When you’re ready to send, touch here.

Press the Home Button when you’re done.

Using FaceTime

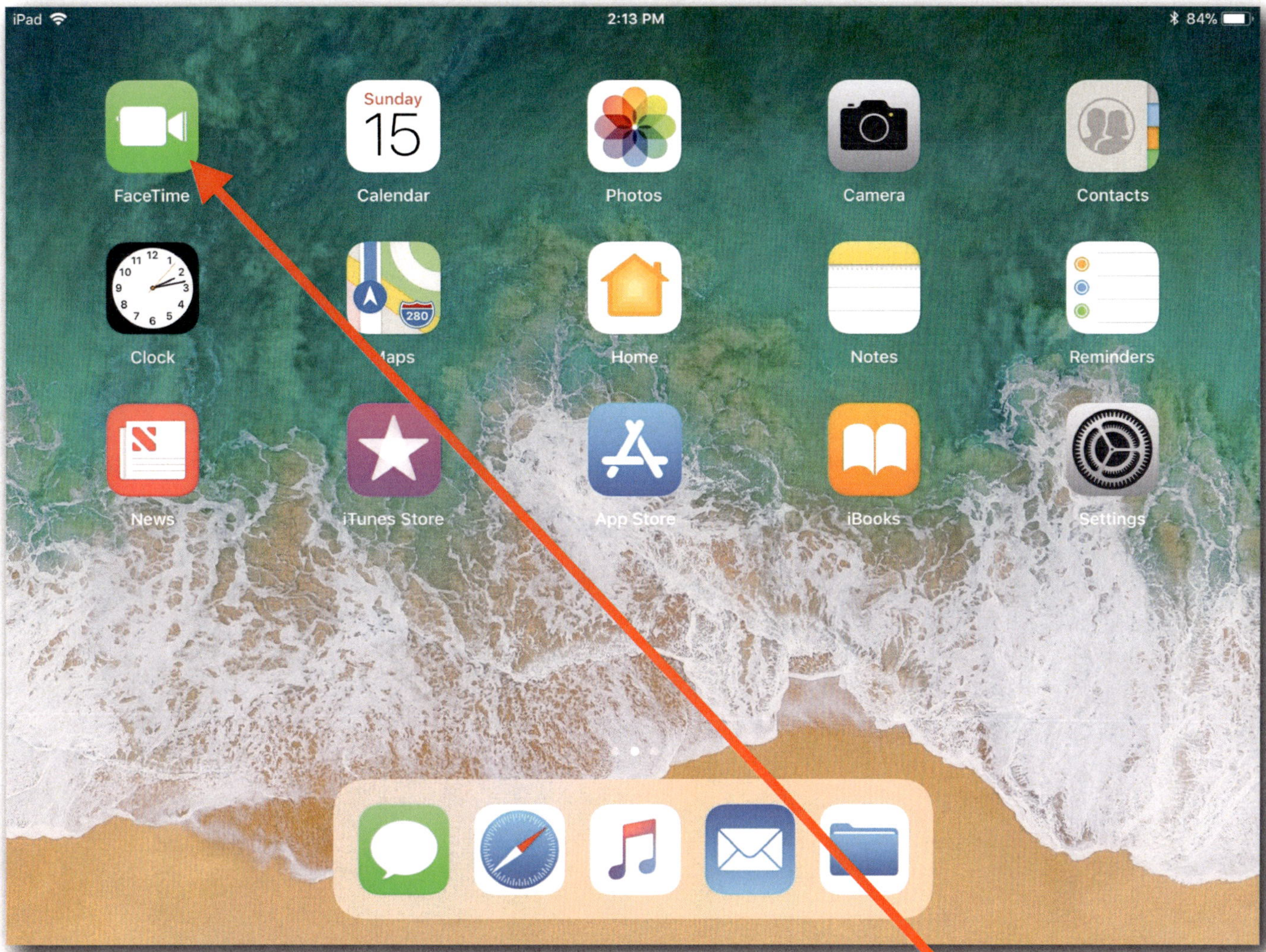

You can use FaceTime to make video or audio calls over the internet, for free.

You must be connected to the internet or WiFi to be able to use FaceTime.

Touch here to begin.

Note: FaceTime calls can only be made to other Apple devices, such as: iPhones, iPads, iPods, or Apple computers.

The first time you use FaceTime you will need to set it up.

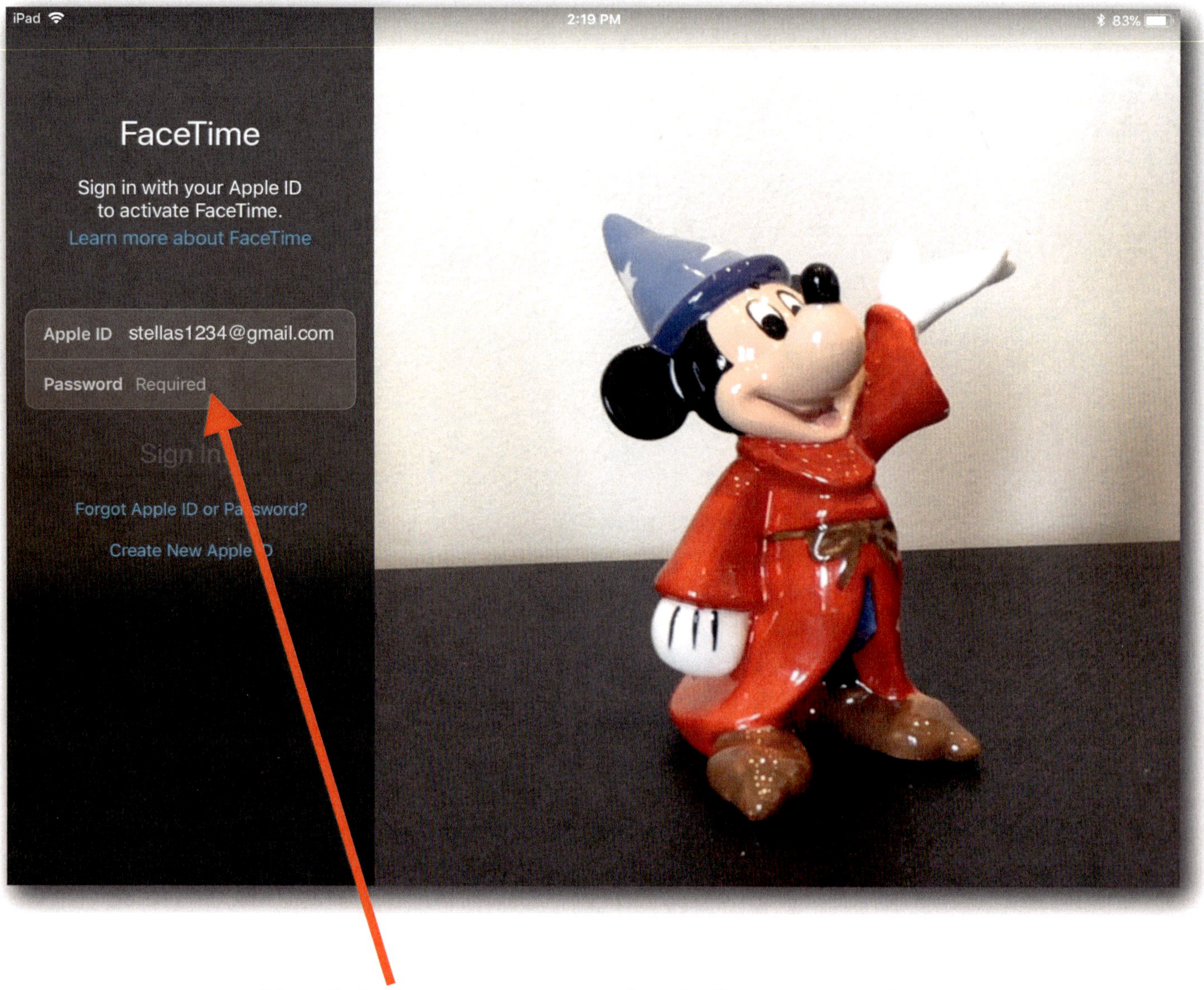

Touch here to enter your Apple ID password.

(For the purpose of this FaceTime demonstration,
I'll be using my figurines from Disney's Sorcerer's Apprentice.)

The spinning wheel means your iPad is signing into FaceTime.

Welcome to FaceTime!

Touch here to begin.

Using the keyboard, type in a name from your contacts, or a phone number, or an email address.

Remember that you can only use phone numbers or email addresses associated with other Apple devices for FaceTime calls.

Touch here to make an audio call. OR Touch here to make a video call.

An audio call will look similar to this.

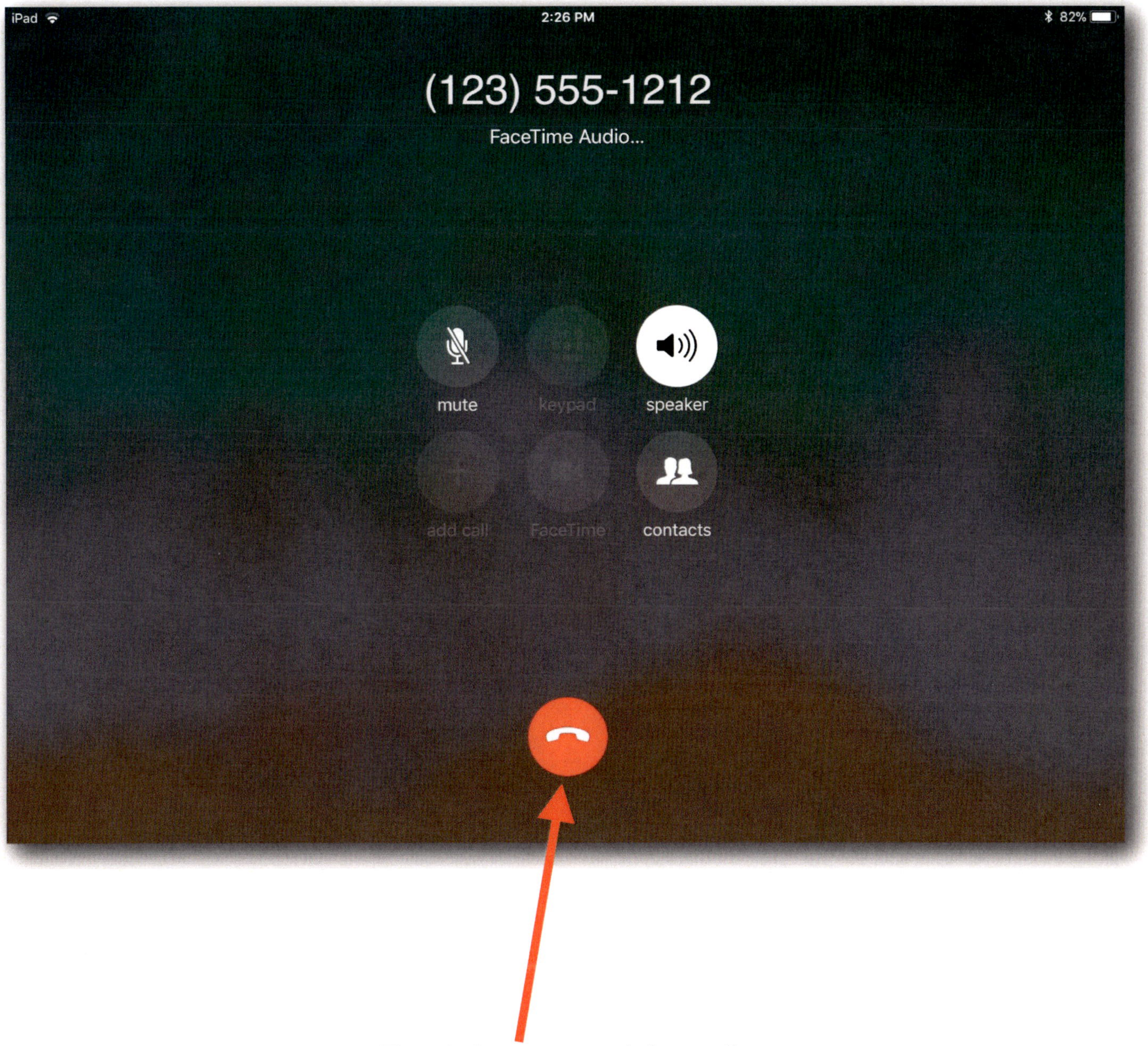

Touch here to end the call.

Press the Home Button when you're done.

A video call will look similar to this.

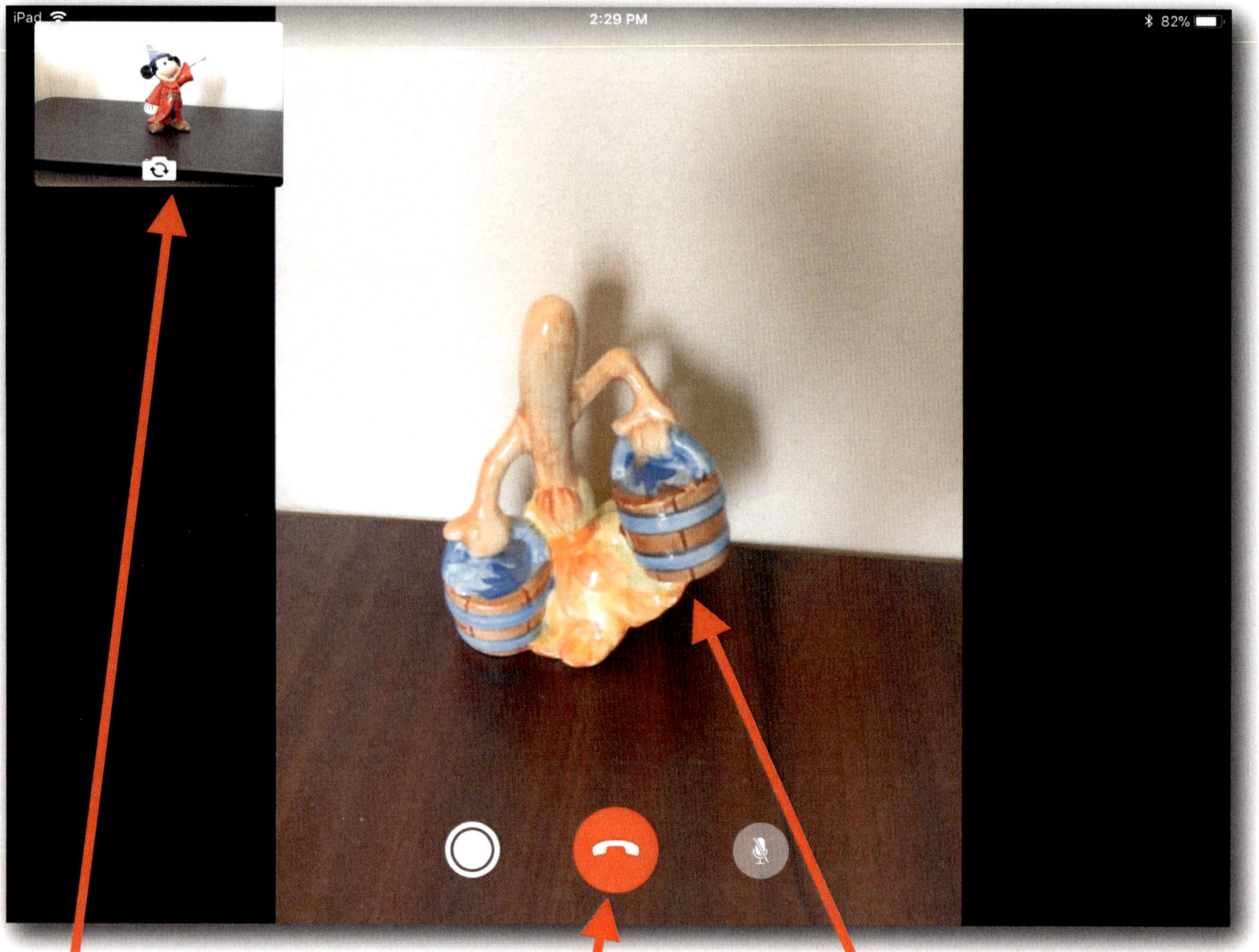

This will be you.

This will be the person you are video-chatting with.

Touch here to end the call.

Press the Home Button when you're done.

Using Your iPad's Camera

Touch here.

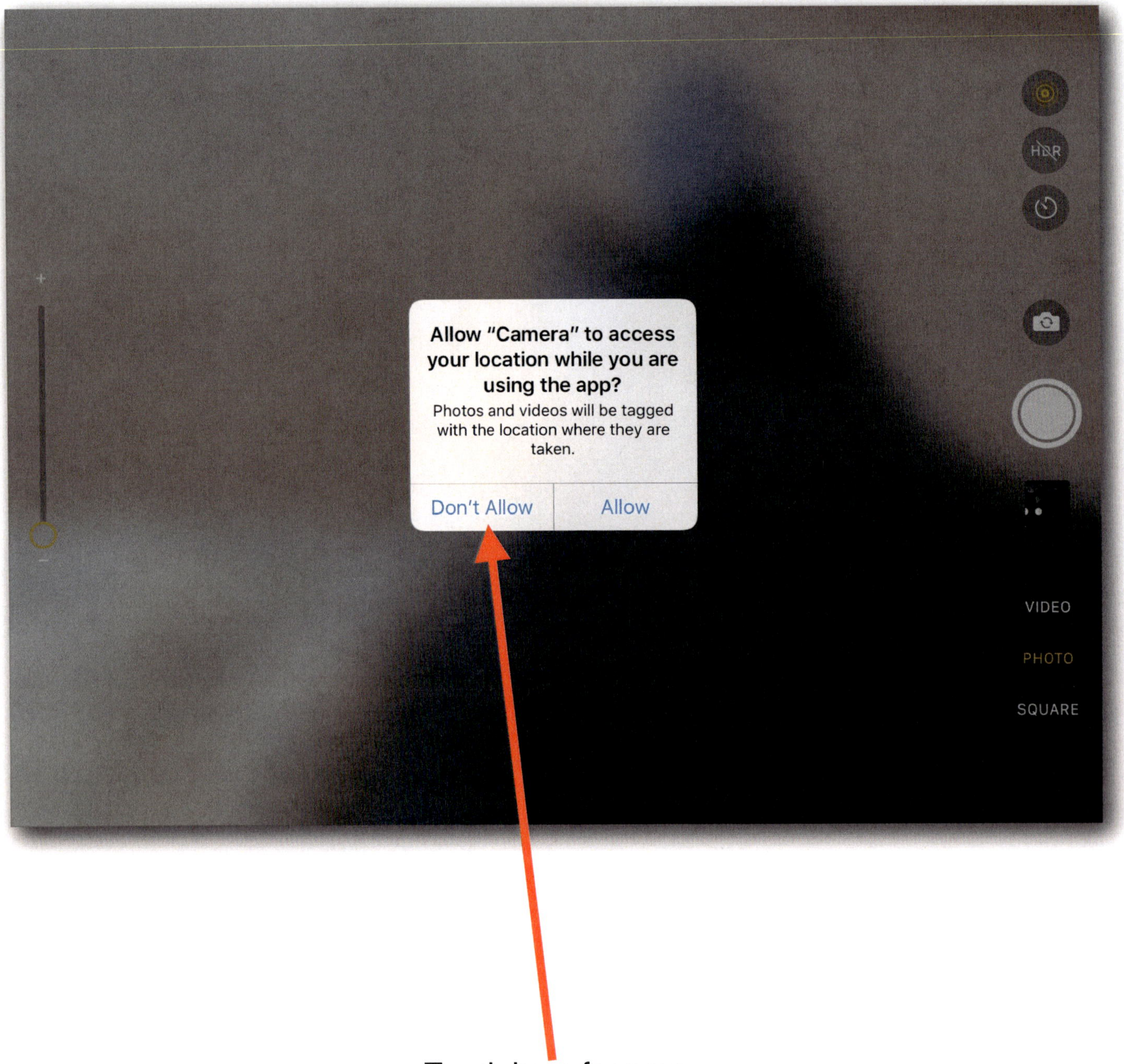
Allow "Camera" to access your location while you are using the app?
Photos and videos will be tagged with the location where they are taken.
Don't Allow
Allow
VIDEO
PHOTO
SQUARE

Touch here for now.

Hold the iPad up to view your subject.

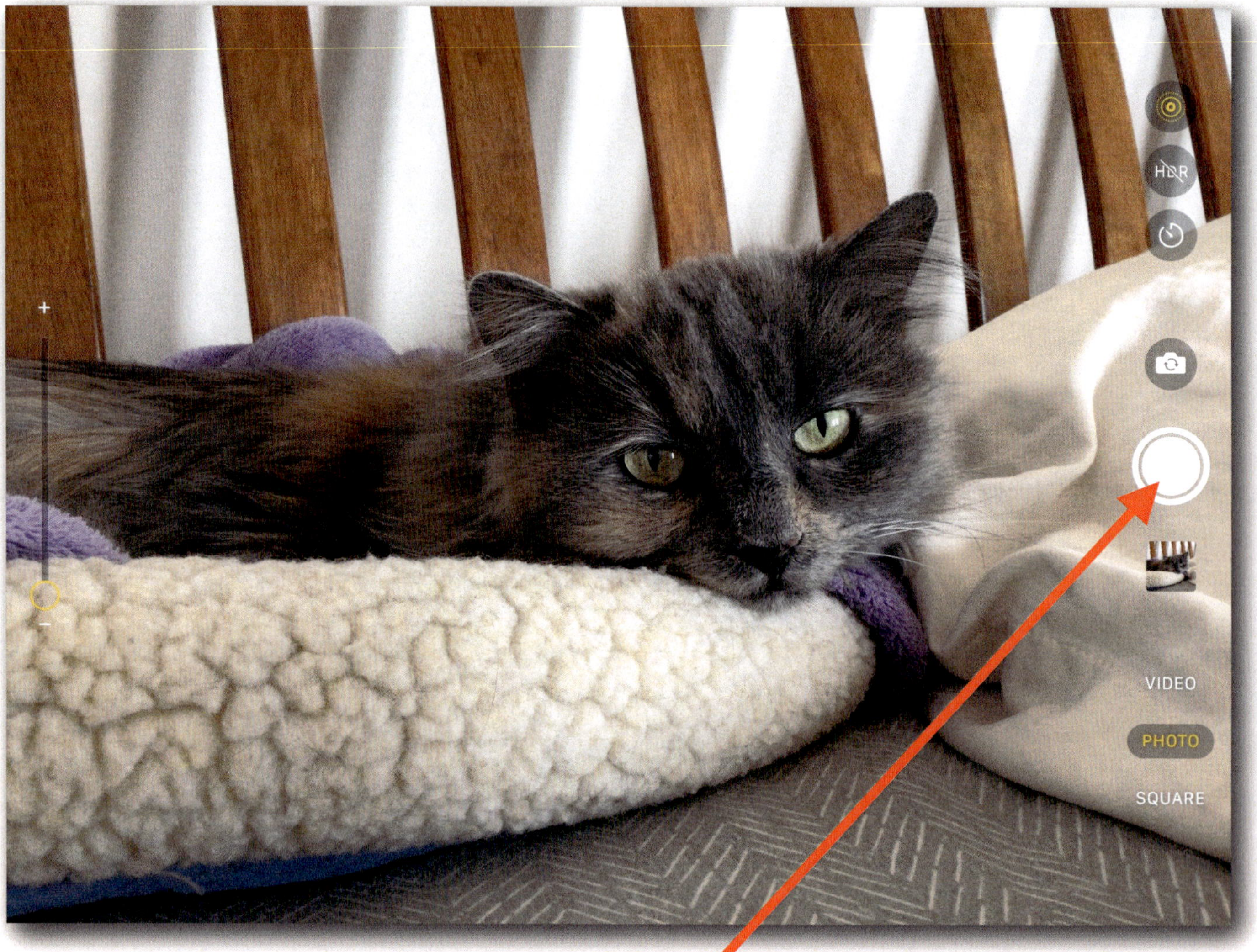

Touch here to take the picture.

The picture will automatically be saved to your Photos Folder.

Press the Home Button when you're done.

Viewing Your Photos

Touch here.

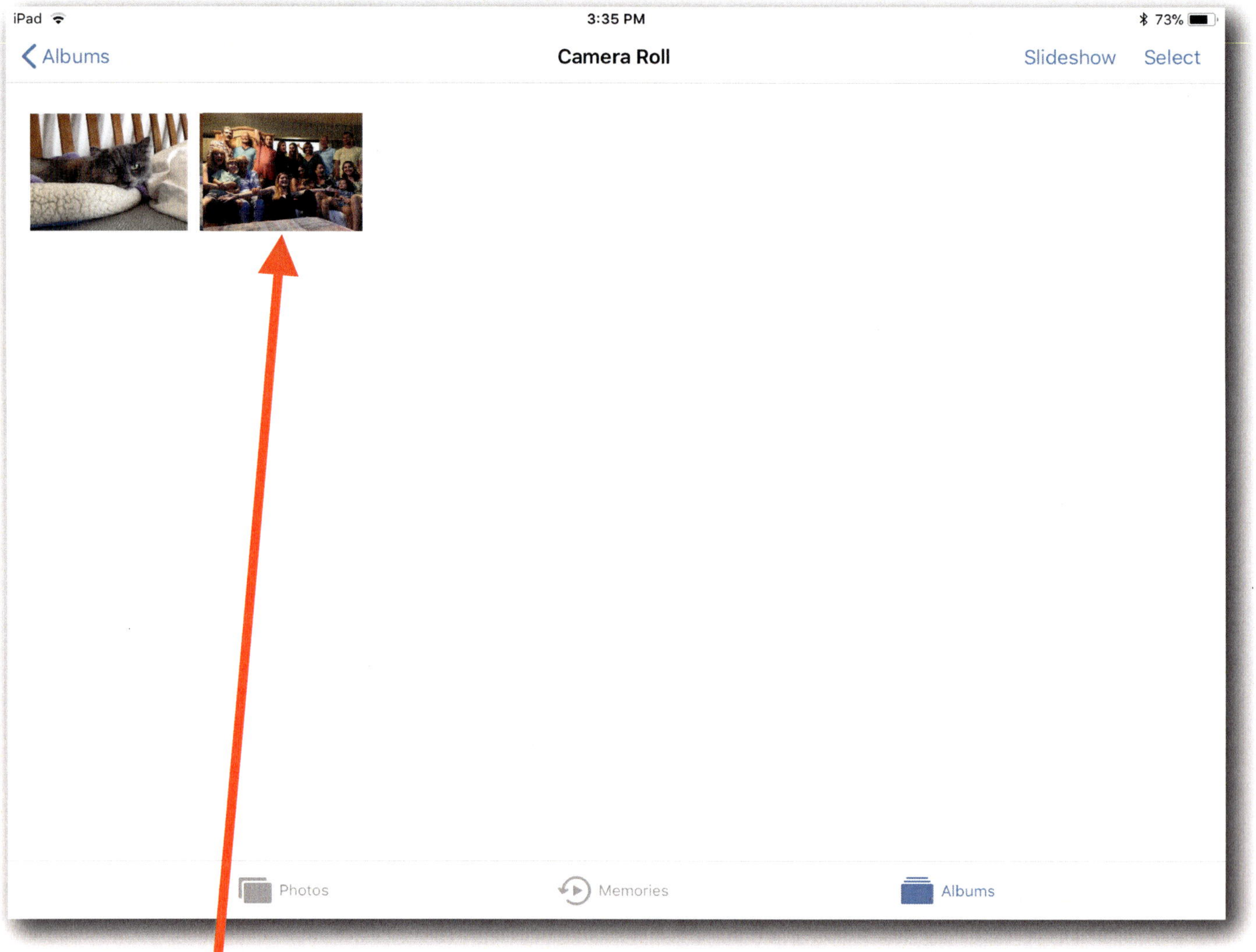

This is your Camera Roll, where "Thumbnails" of all of your photos are stored.

Touch a photo you wish to view.

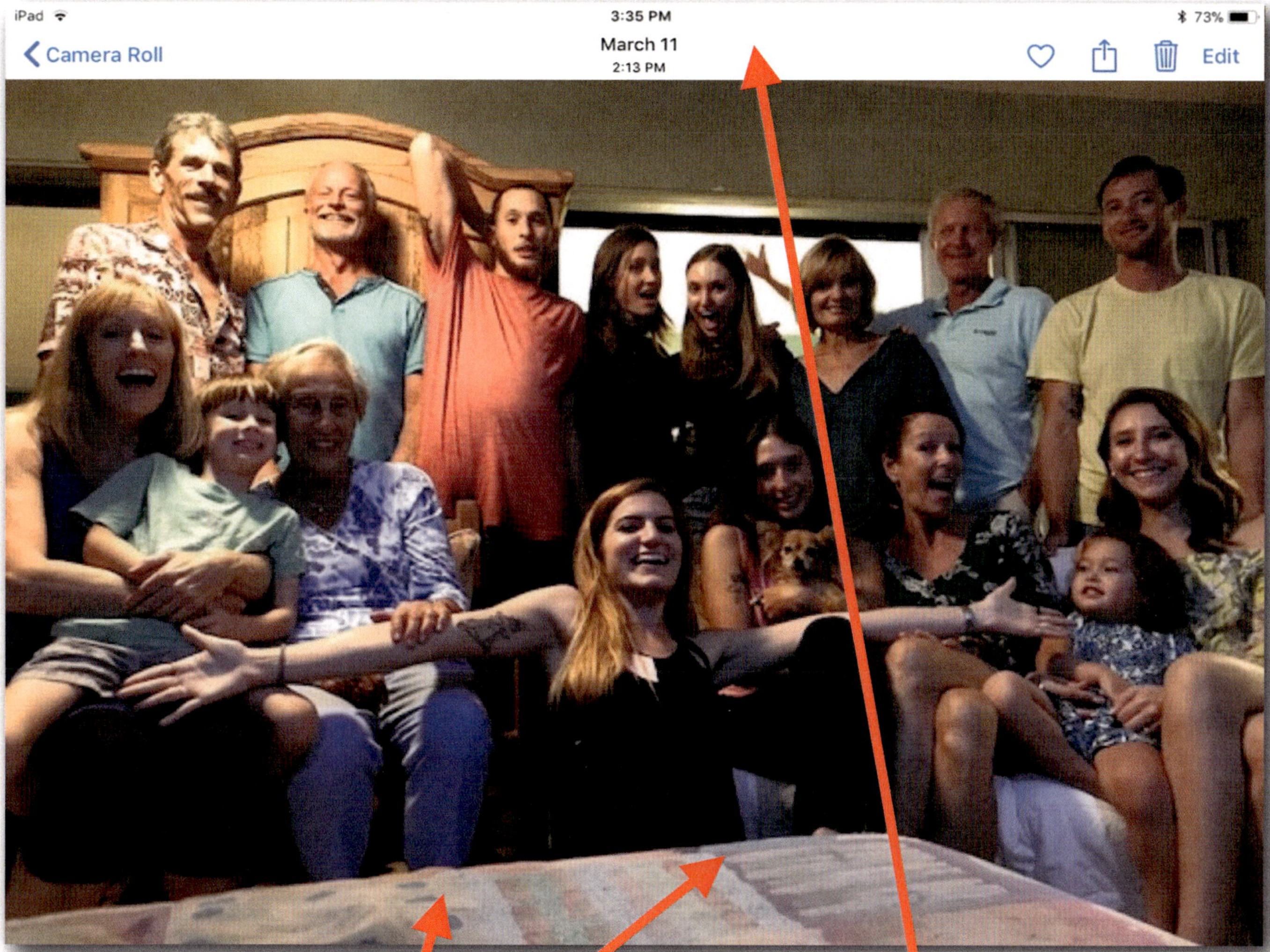

Touch anywhere in the photo to make this status bar disappear.

Touch anywhere in in the photo again to bring back the status bar.

Touch here to return to the Camera Roll.

Press the Home Button when you're done.

Reading The News On Your News App

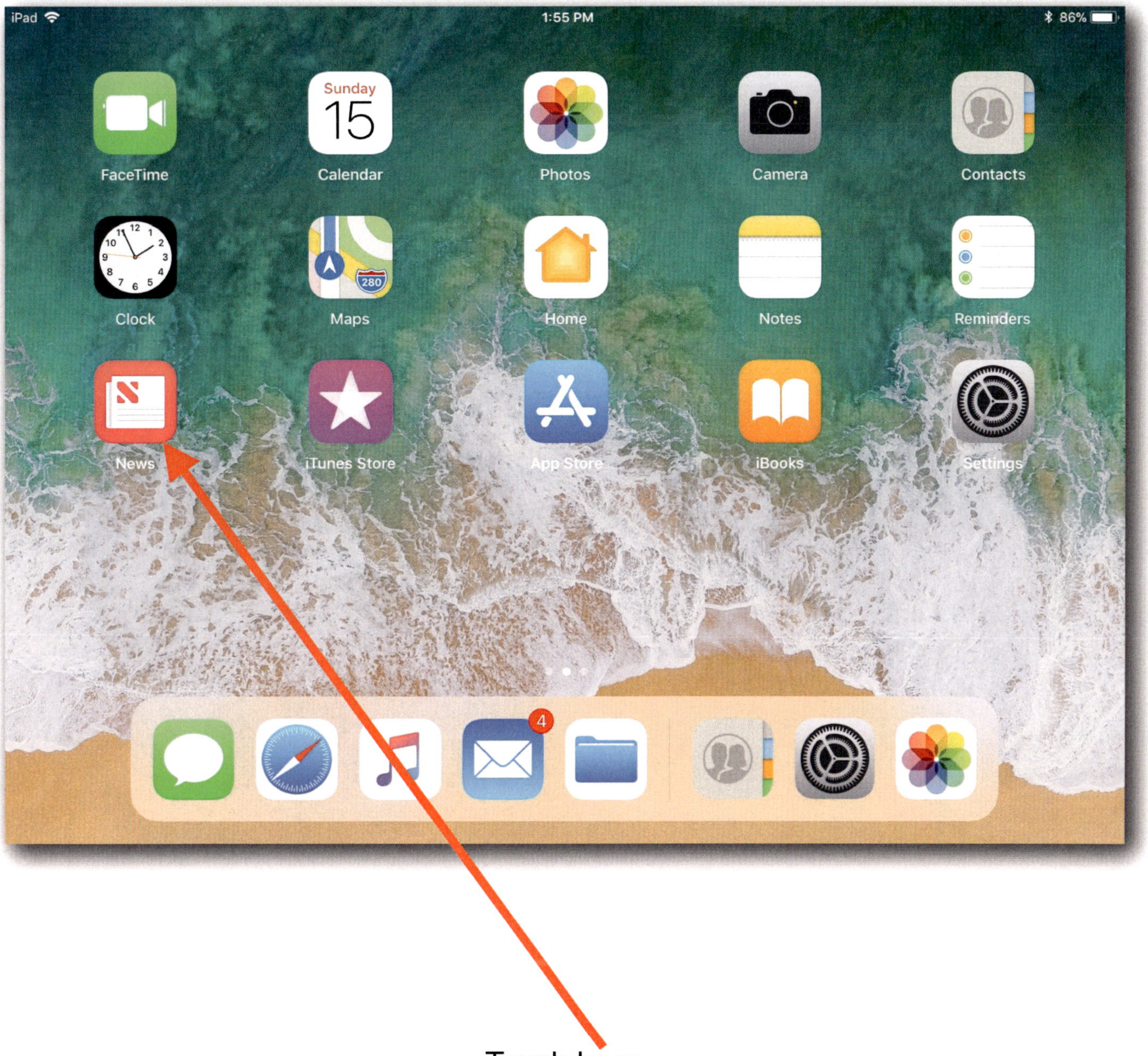

Touch here.

iPad
2:05 PM
85%
SUNDAY
APRIL 15
Siri Suggestions
Siri suggests stories, channels, and to
Safari and app usage. You can chang
News and Privacy
TOP STORIES
Allow "News" to access your location while you are using the app?
Your location is used to deliver locally relevant information such as search results and weather.
Don't Allow
Allow
For You
Spotlight
Following
Search
Saved

Touch here for now.

Scroll down the page to view headlines and touch any article you wish to read.

Press the Home Button when you're done.

Bonus Chapter: iPhone Tutorial

Answering An Incoming Call

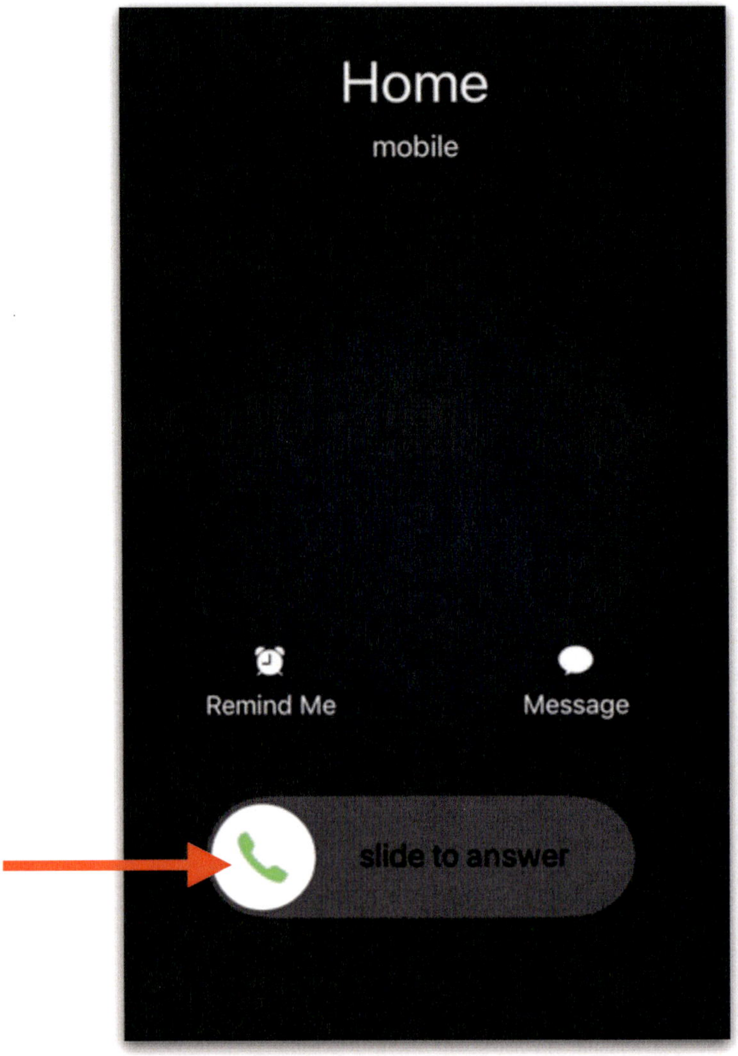

To answer an incoming call that looks like this, slide the little phone icon to the right.

An incoming call will look like this when your phone is not in use at the time of the call.

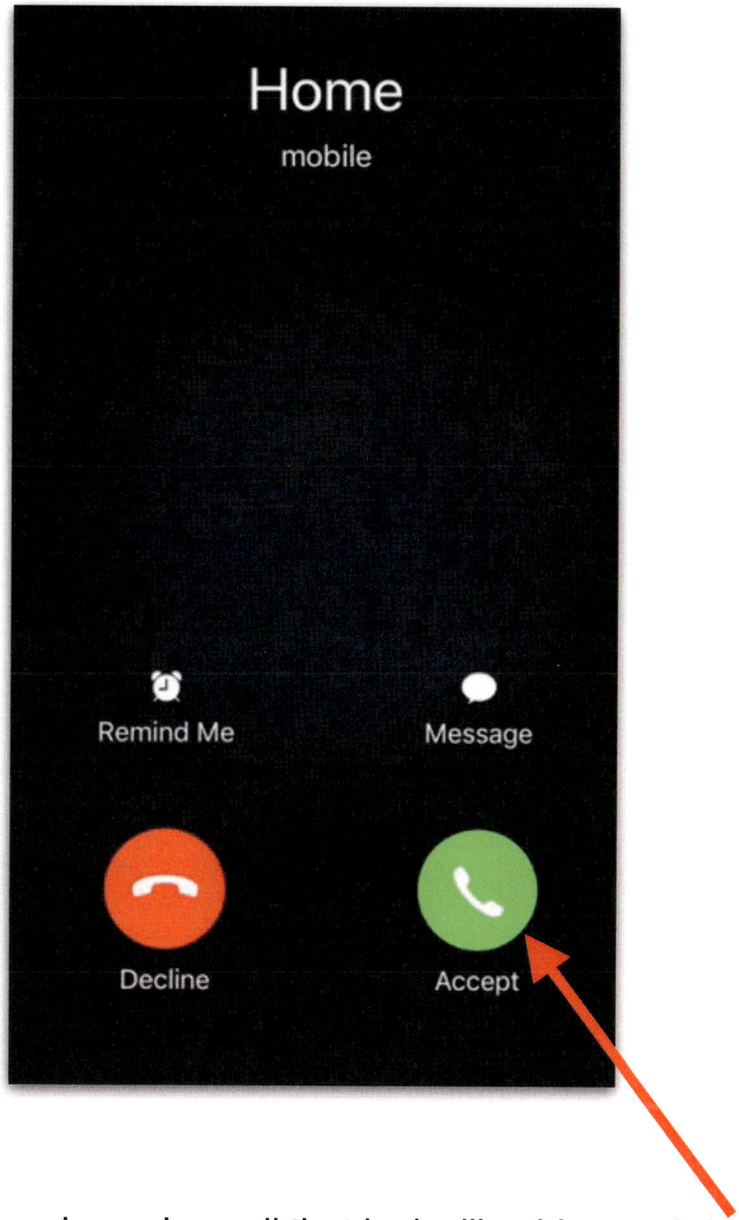

To answer an incoming call that looks like this, touch here.

An incoming call will look like this when you are using your phone for something other than a call (such as checking email or text messaging) at the time of the incoming call.

Ending A Call

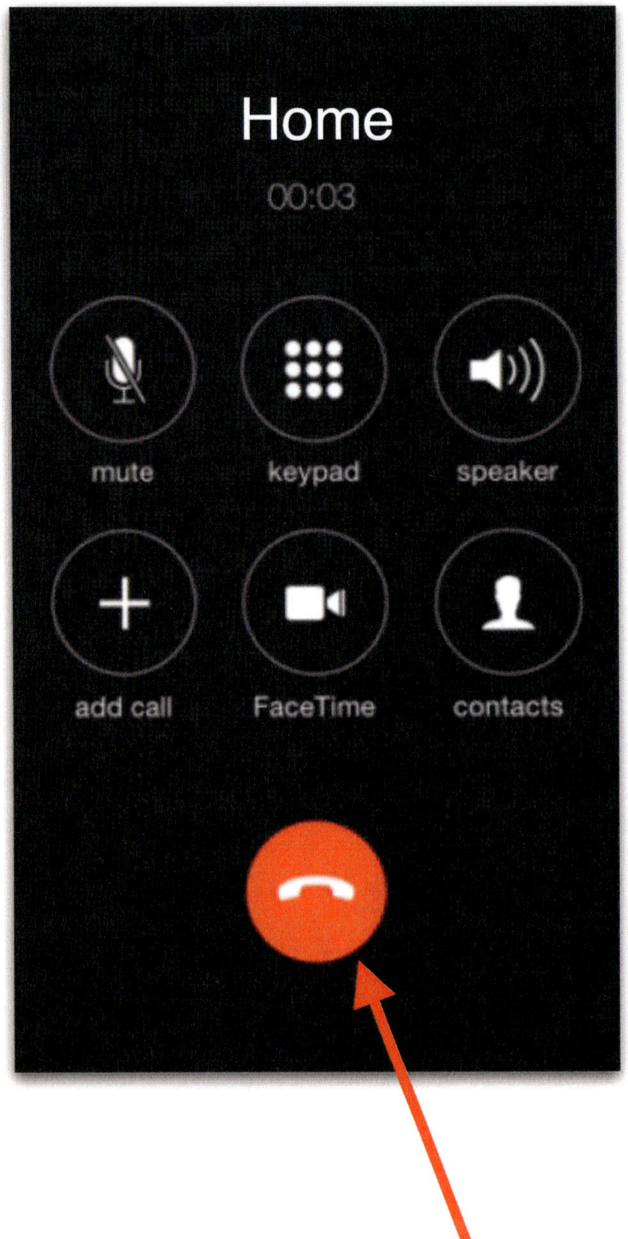

To end a call, touch here.

Press the Home Button when you're done.

Using Your Phone Features:

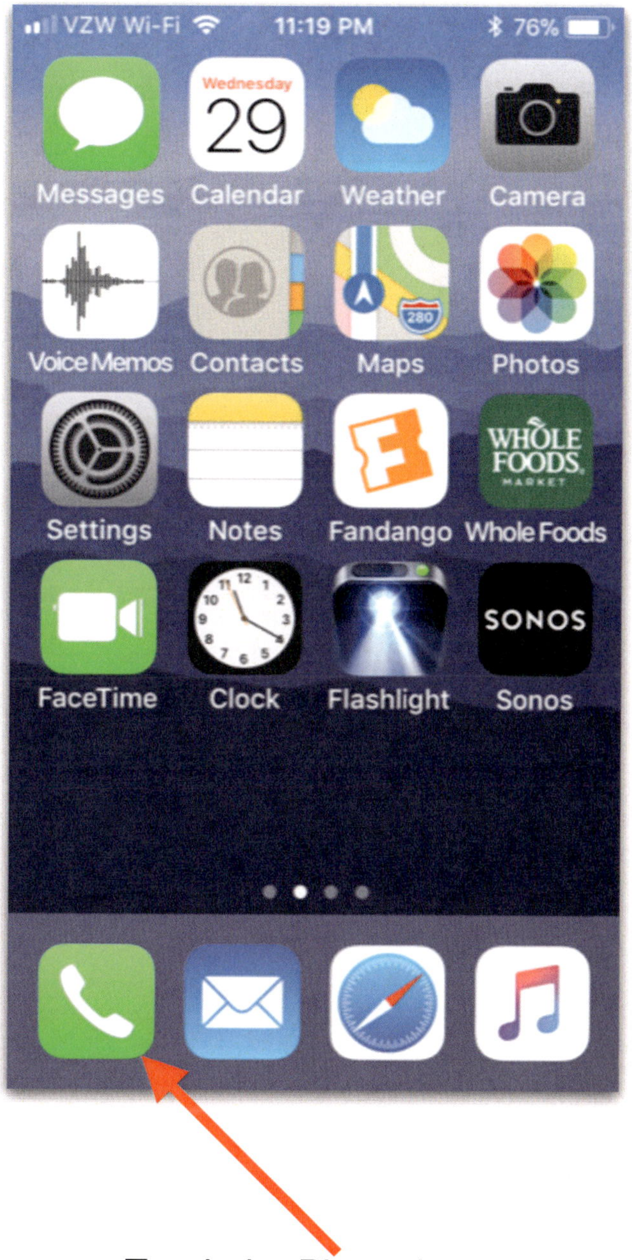

Touch the Phone icon.

Listening To Voicemail

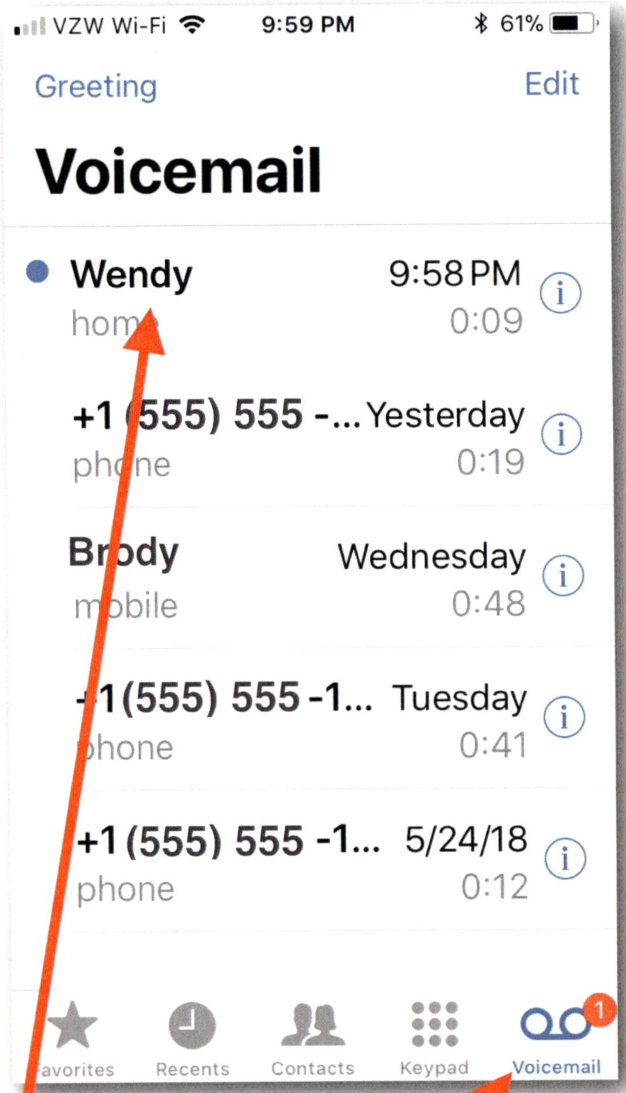

Touch here to check voicemail.

Then…

Touch the voicemail you'd like to listen to.

Note: A blue dot indicates an unheard voicemail.

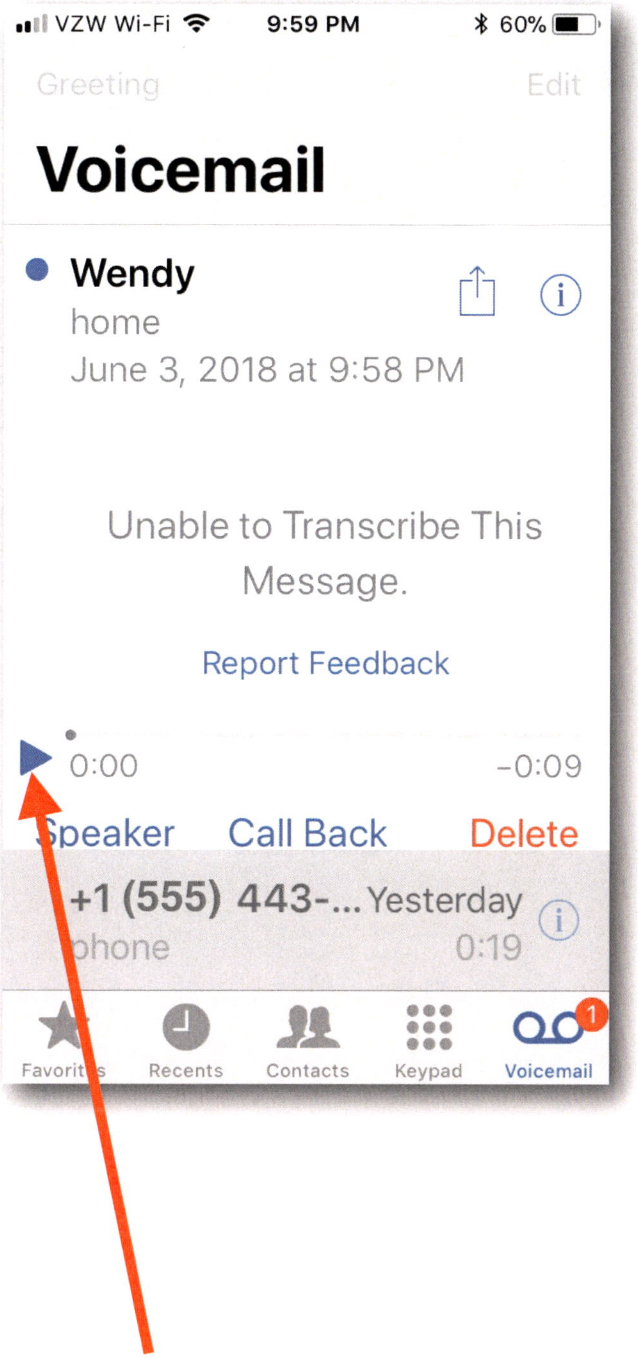

Touch here to listen to your voicemail.

Dialing A Number

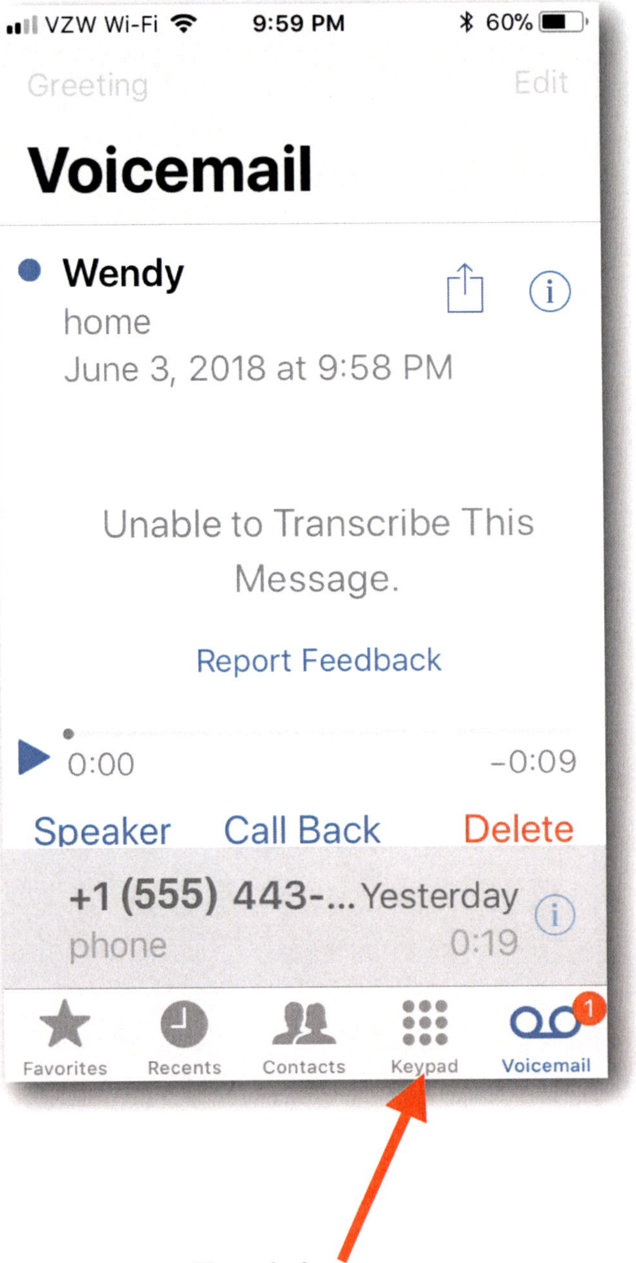

Touch here.

Enter the phone number on the keypad, then touch here to connect.

Using Your Contacts

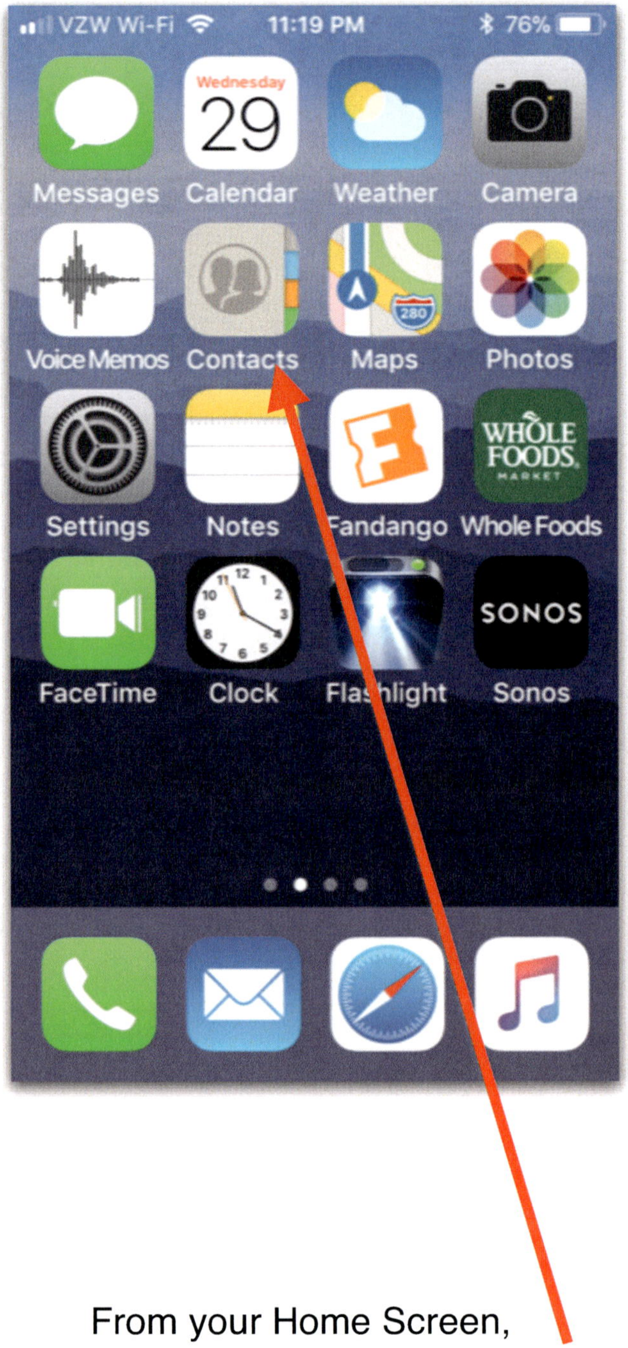

Touch here. OR From your Home Screen, touch here.

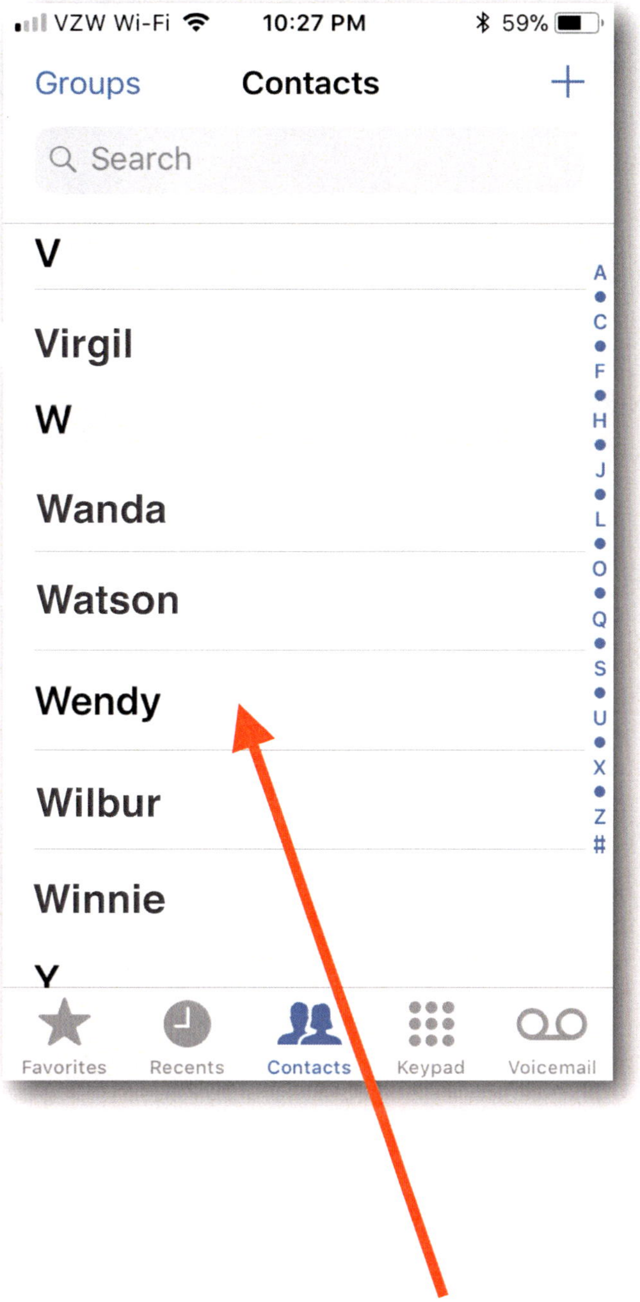

Touch your chosen contact.

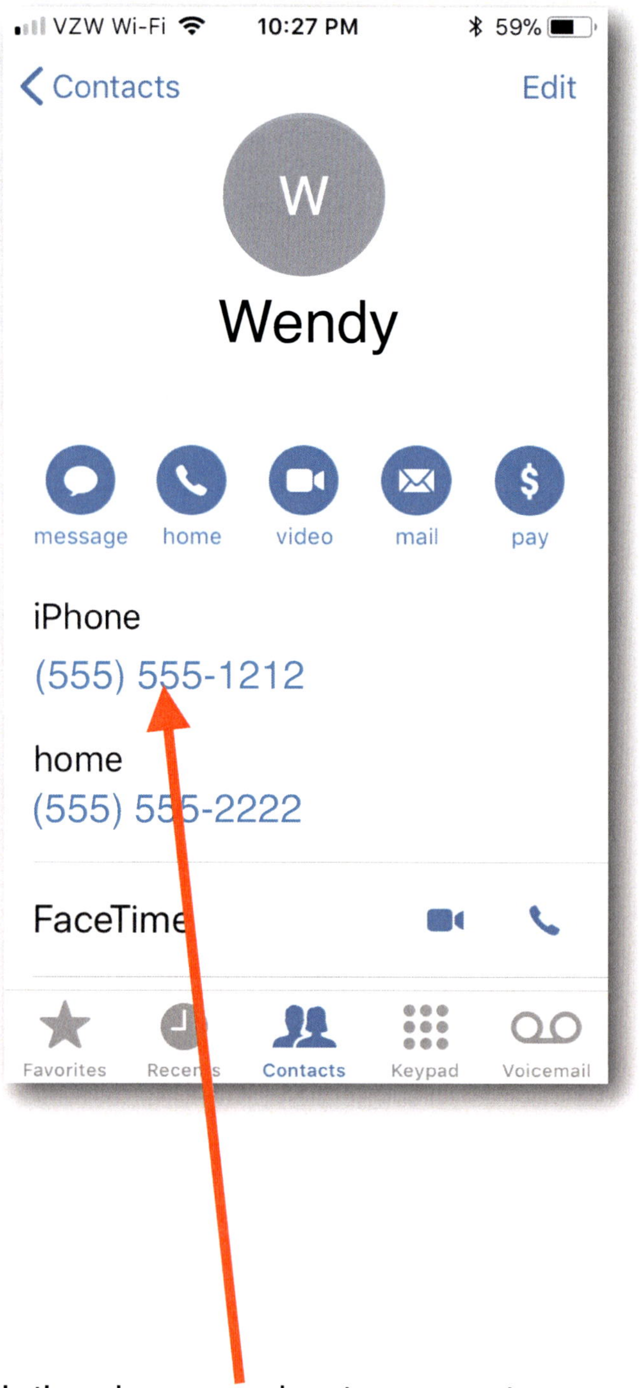

Touch the phone number to connect your call.

Checking Recent Calls

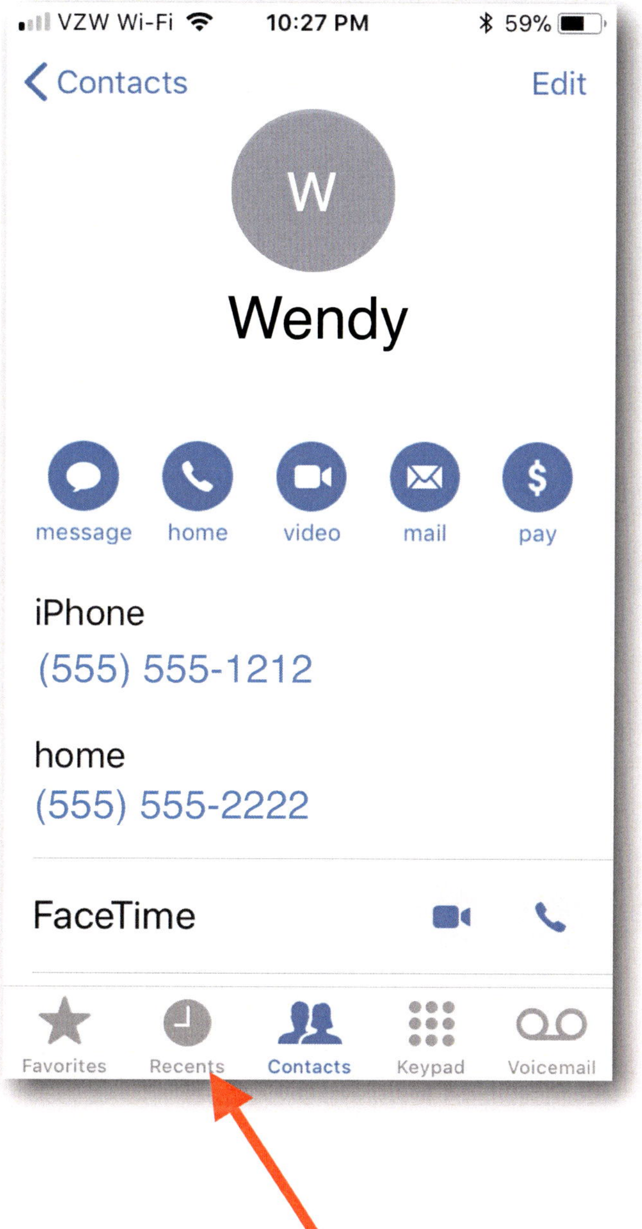

Touch here.

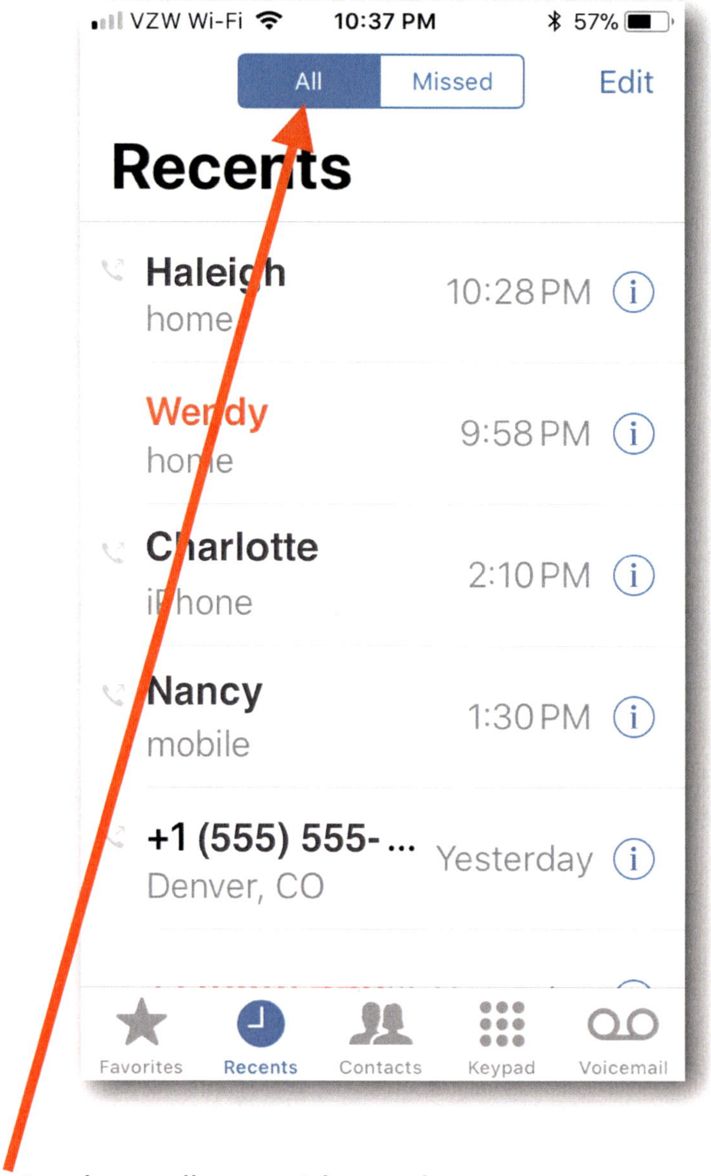

Touch here to show all recent incoming
and outgoing calls.

Red type indicates a missed incoming call.

Black type indicates a call you made out or
a call you successfully received.

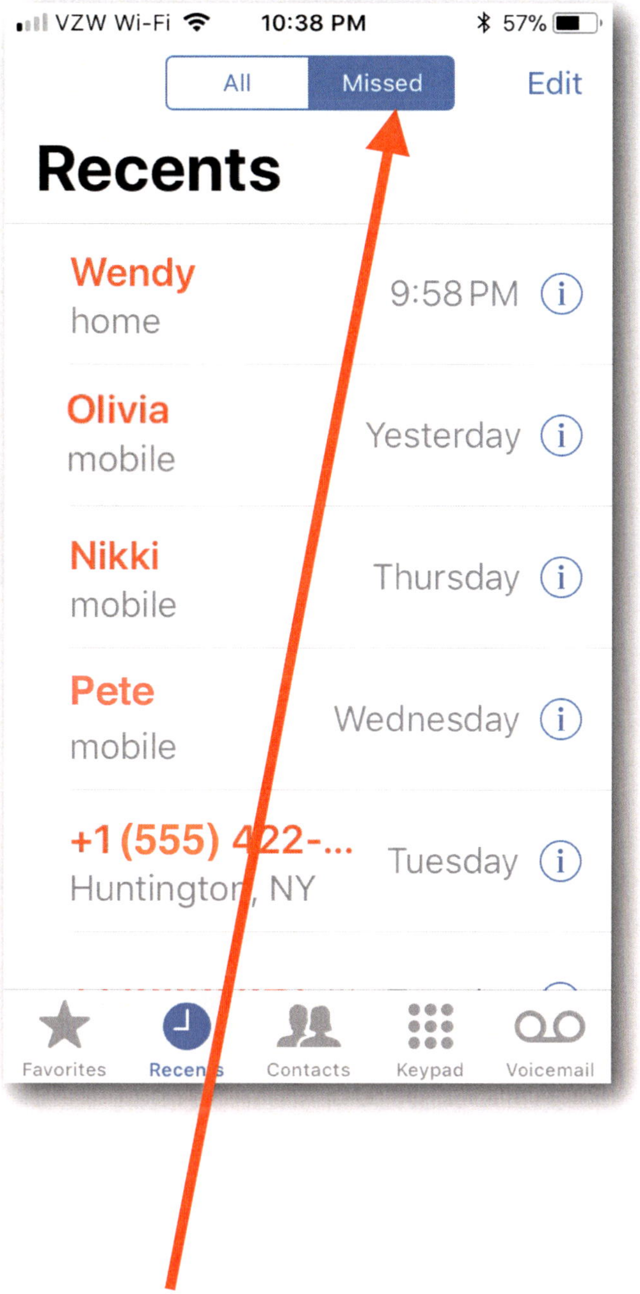

Touch here to show only missed incoming calls.

Using Your Favorites File

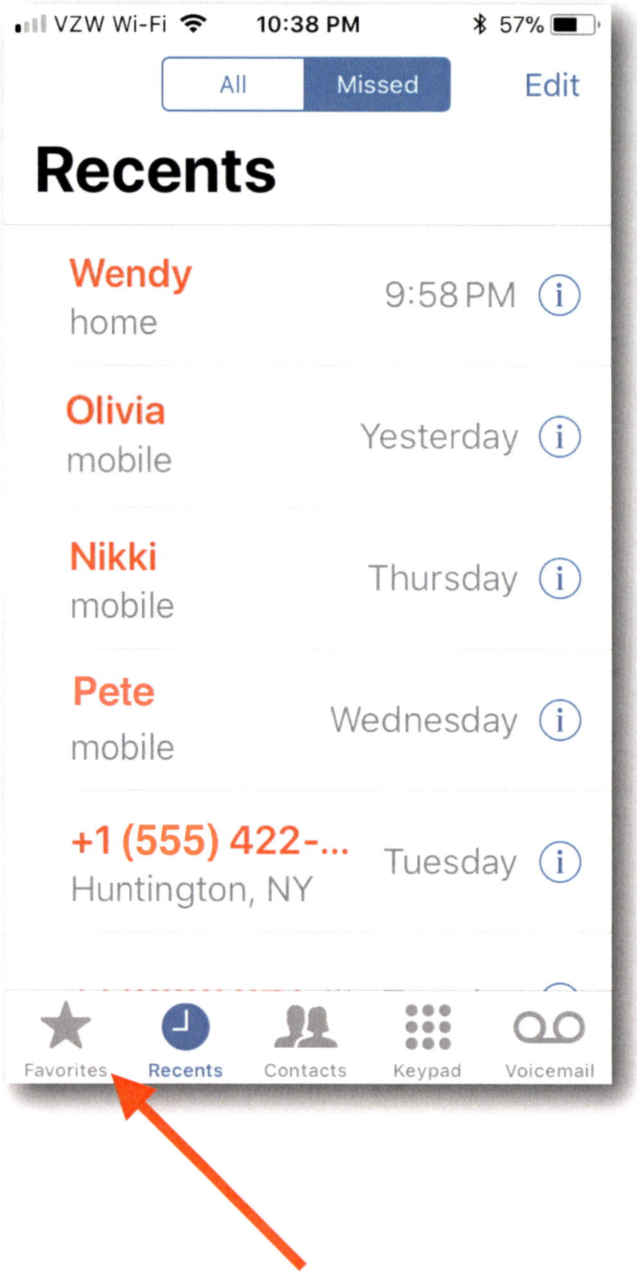

Touch here.

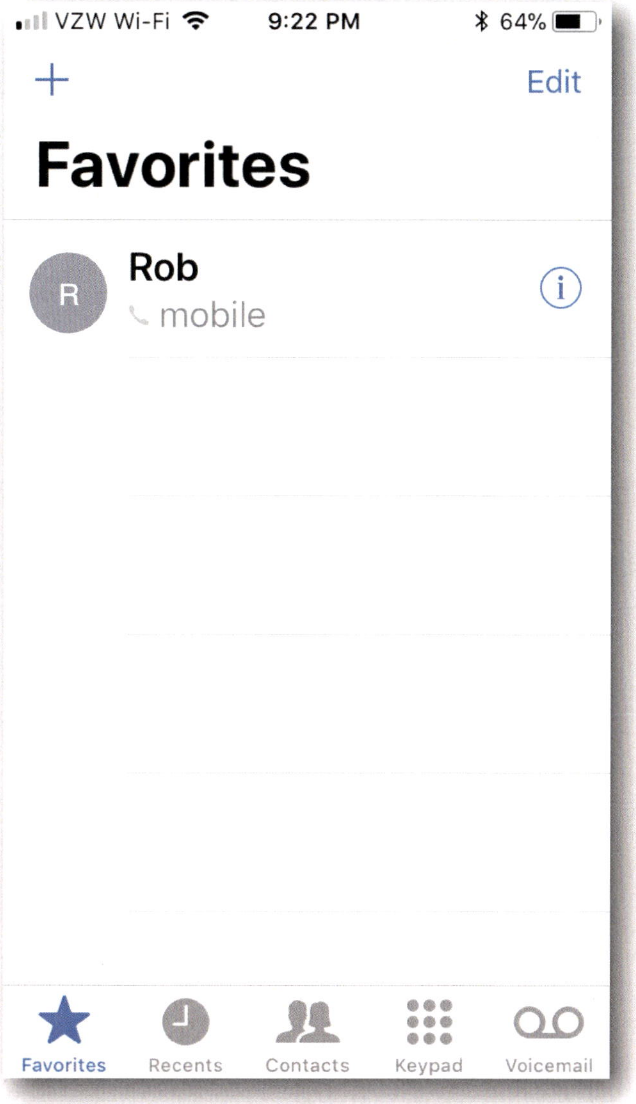

Your Favorites file allows for a shortcut when calling a number, or numbers, you frequently use.

(I only keep one contact in my Favorites file so that I can make the call quickly. This is a shortcut because I have many people in my Contacts and prefer not to scroll through the list.)

If you'd like to add a contact to your Favorites file, follow the steps on the next few pages.

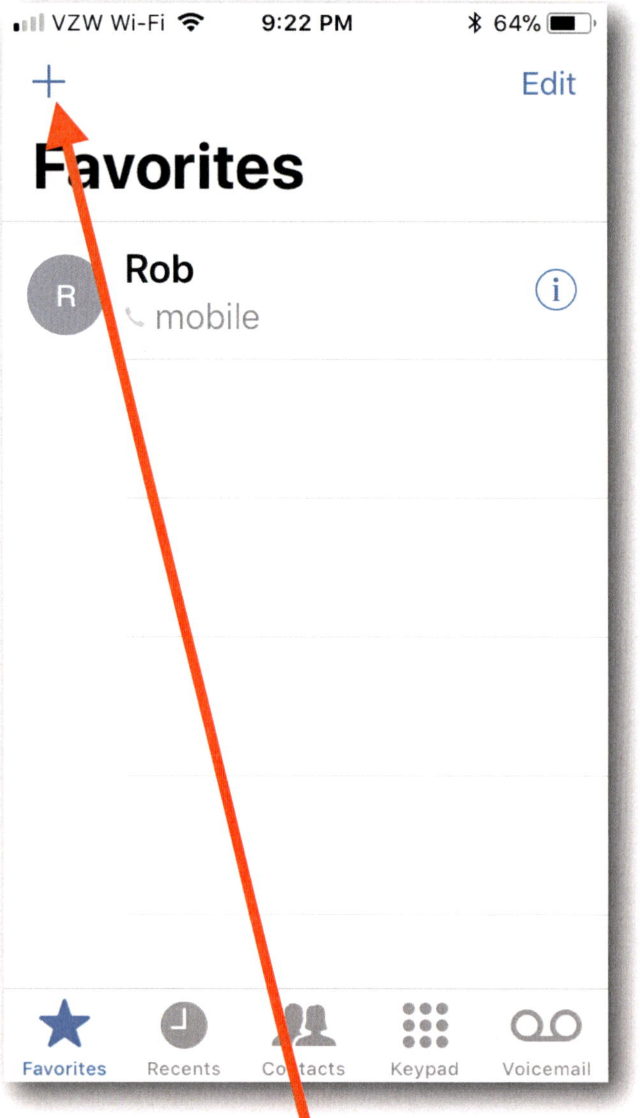
VZW Wi-Fi
9:22 PM
64%
Edit
Favorites
Rob
mobile
Favorites
Recents
Contacts
Keypad
Voicemail

Touch here.

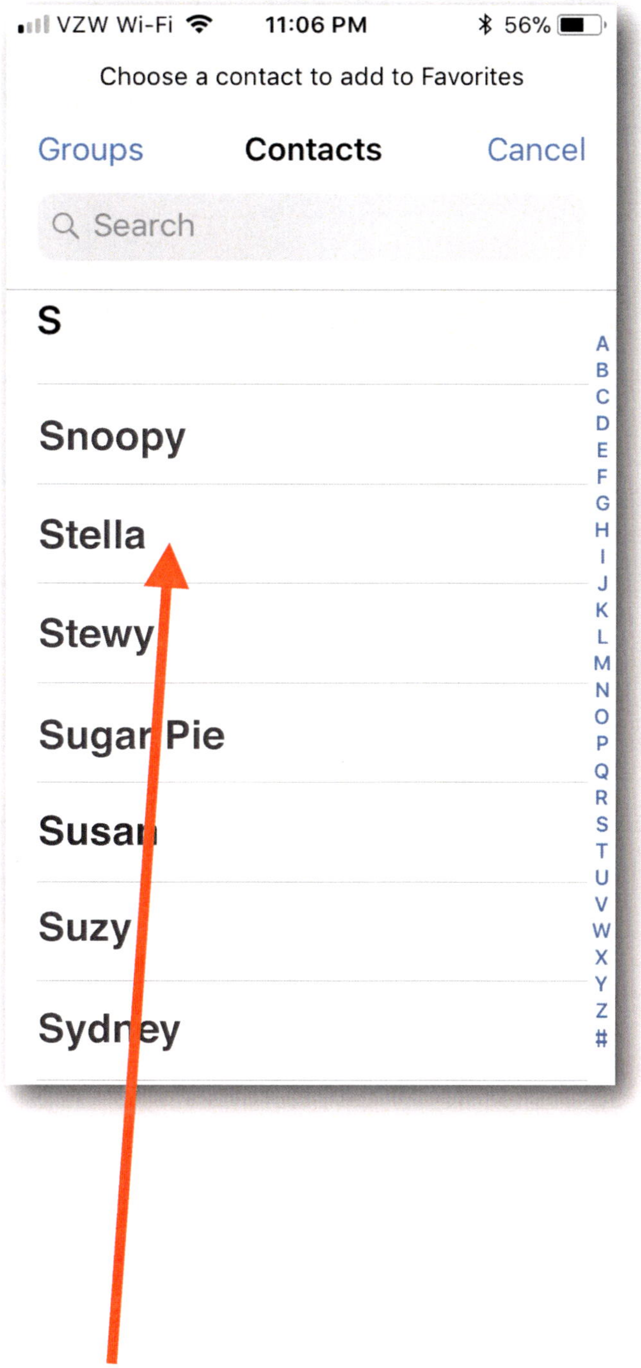

Touch the contact you'd like to add to your Favorites file.

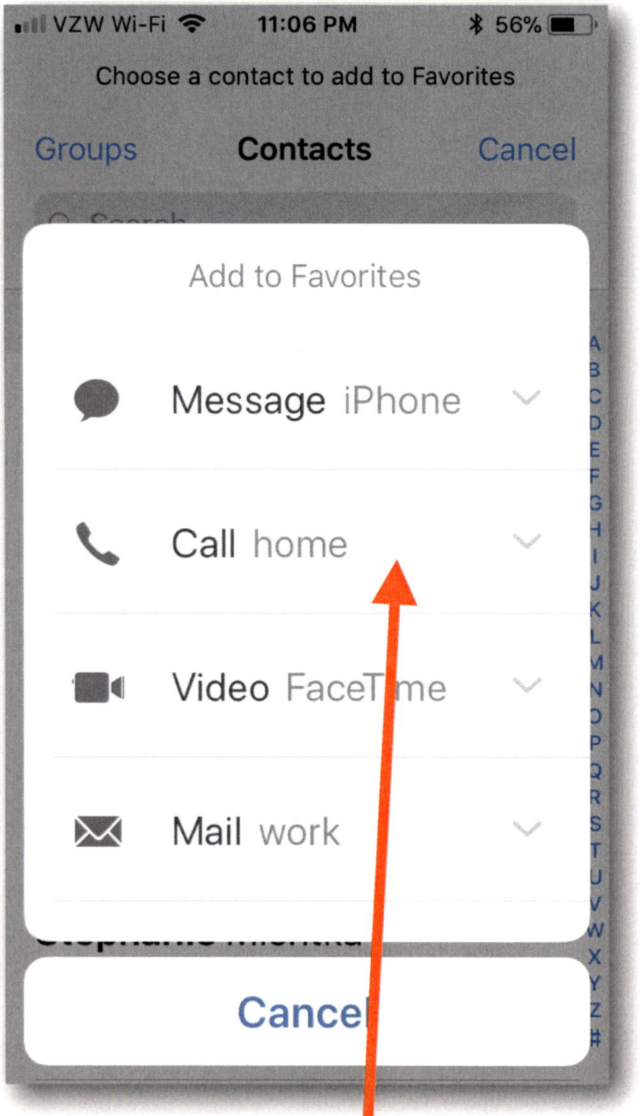
VZW Wi-Fi
11:06 PM
56%
Choose a contact to add to Favorites
Groups
Contacts
Cancel
Add to Favorites
Message iPhone
Call home
Video FaceTime
Mail work
Cancel

Touch here.

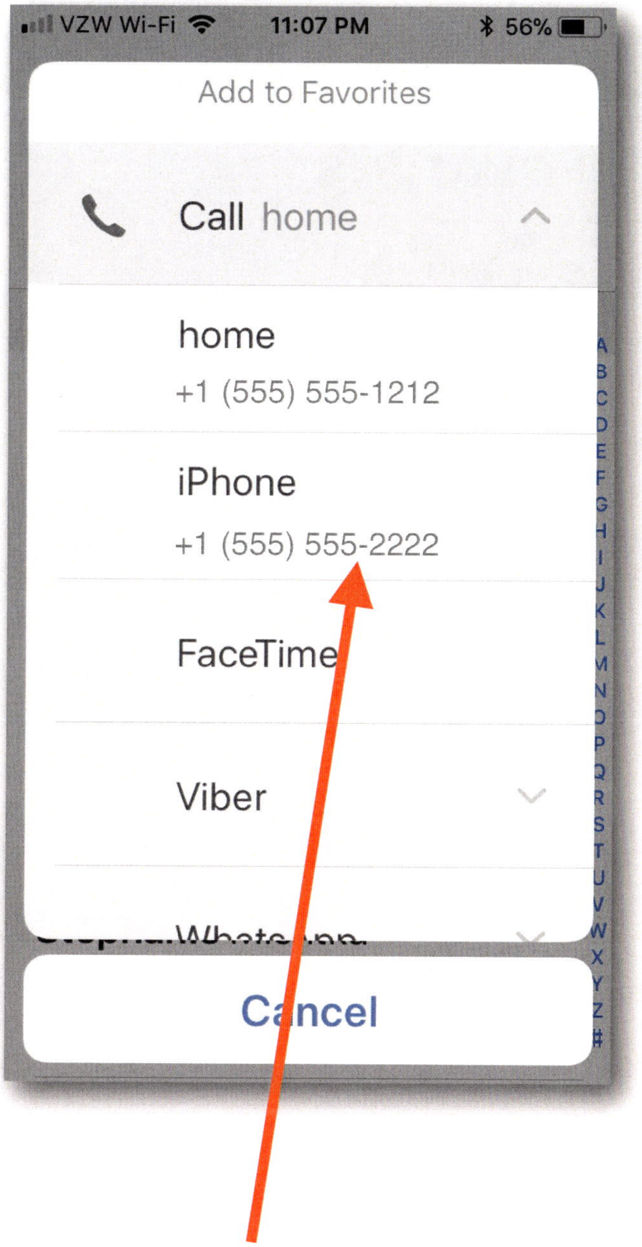

Touch the number you'd like to use.

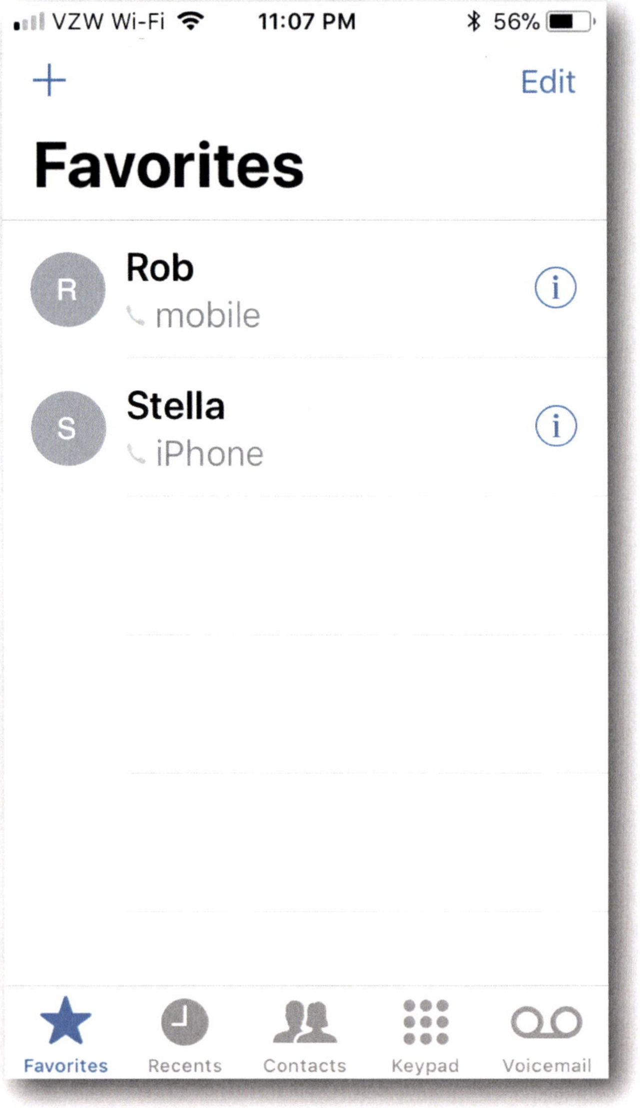

Voila! You have added a contact to your Favorites file!

Helpful Hints and Troubleshooting

If your screen looks similar to this,
press the Home Button to get back to the home page.

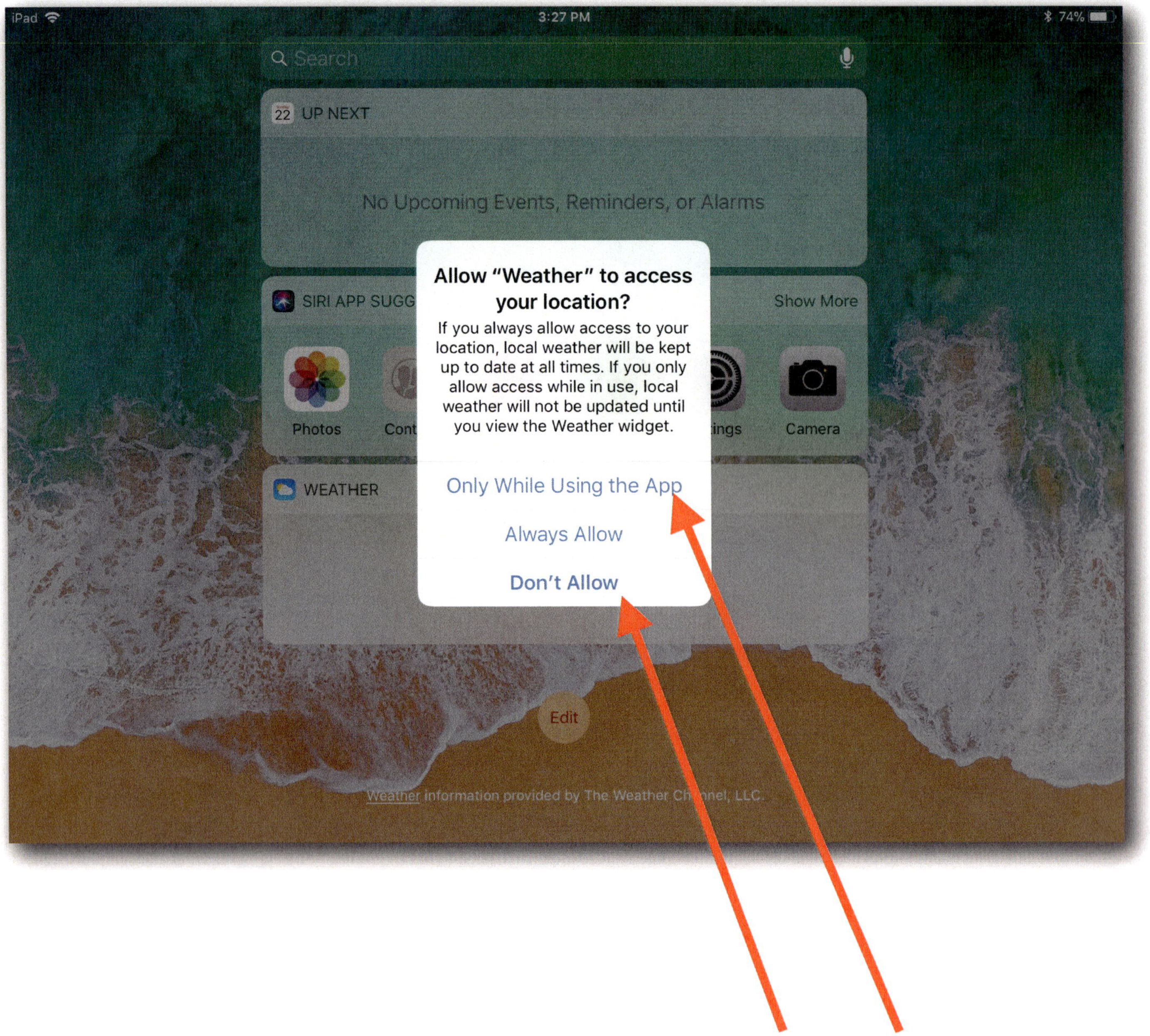
iPad
3:27 PM
74%
Search
UP NEXT
No Upcoming Events, Reminders, or Alarms
SIRI APP SUGG
Show More
Photos
Camera
WEATHER
Allow "Weather" to access your location?
If you always allow access to your location, local weather will be kept up to date at all times. If you only allow access while in use, local weather will not be updated until you view the Weather widget.
Only While Using the App
Always Allow
Don't Allow
Edit

If this pops up in the middle of your screen, touch here or here, for now.

If your screen looks similar to this,
press the Home Button to return to the home page.

When you first wake up your iPad there may be alerts, such as this one.

Press the Home Button to return to the home page.

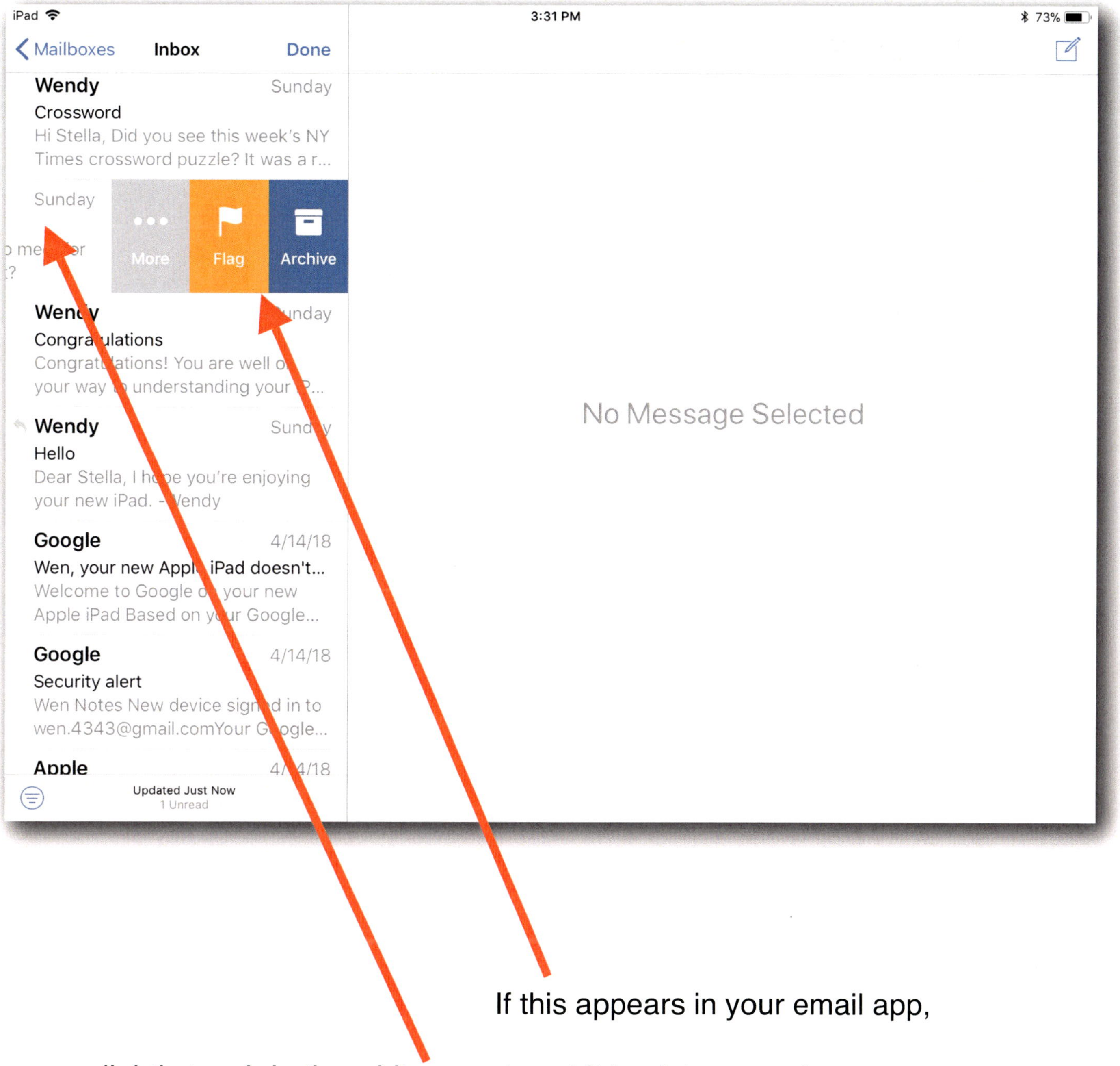

If this appears in your email app,

lightly touch in the white area to set it back to normal.

This is the main page of the home screen.

To turn to page two of the home screen, swipe to the left.

This is page two of the home screen.

To return to page one, swipe to the right or press the Home Button.

"Jiggly" Mode

Icons will go into "Jiggly Mode" if you touch and hold any icon for three seconds or more.

To exit "Jiggly Mode," press the Home Button.

Note: Do not touch any of the X's in the top left corner of any icon in "Jiggly Mode" or you will delete that app from your iPad.

After pressing the Home Button, your iPad Home Page will return to normal.

iPad Becomes Sluggish

If your iPad becomes sluggish or slow to respond, it is an indication that the iPad needs to be shut down completely and restarted (not to be confused with putting the iPad into Sleep Mode). The process of shutting down and restarting will refresh the iPad.

To shut down the iPad, follow the instructions on page 5.

Congratulations!

You have reached the end of this book!

After you've practiced and mastered these beginning steps, you may find that you would like to learn more. Exploring more iPad features should be easier, now that the absolute basics have been mastered.

There are many books for seniors out there, which go into much more detail and explore more iPad features than I did in this book.

My objective for this book was only to get you started.

I wish you much success on your iPad journey!

Made in United States
Orlando, FL
04 November 2026

90201163R00121